MASTITIS CONTROL IN DAIRY HERDS
An Illustrated and Practical Guide

MASTITIS CONTROL IN DAIRY HERDS
An Illustrated and Practical Guide

Roger Blowey
B.Sc., B.V.Sc., F.R.C.V.S.

Peter Edmondson
M.V.B., Cert. CHP, F.R.C.V.S.

Farming Press

First published 1995

ISBN 0 85236 314 1

A catalogue record for this book is available
from the British Library

Published by Farming Press Books
Miller Freeman Professional Ltd
Wharfedale Road, Ipswich IP1 4LG, United Kingdom

Distributed in North America
by Diamond Farm Enterprises,
Box 537, Alexandria Bay, NY 13607, USA

Line drawings by Jane Upton

Cover design by Andrew Thistlethwaite

Typeset by Galleon Typesetting
Printed and bound in Great Britain by Bath Press

Contents

Dedicated to Norma and Parveen

Acknowledgements

This book is based on notes we prepared for our seminars on mastitis control and quality milk production. We would like to thank those involved in the organisation of the seminars: Catherine Girdler and Susan Evans for typing the manuscript and Jenny Davis for advice on microbiology. We are also grateful to the many authors whose data we have quoted, and to the farmers in our respective veterinary practices who have allowed us onto their farms to train students and veterinary surgeons and to take numerous photographs. Special thanks go to Jane Upton for bringing the text to life with such excellent artwork and to our respective wives, Norma and Parveen, for their patience, tolerance and support during the preparation of the manuscript.

ROGER BLOWEY, *B.Sc., B.V.Sc., F.R.C.V.S.*

PETER EDMONDSON, *M.V.B., Cert.CHP, F.R.C.V.S.*

CHAPTER ONE

Introduction

The aim of this book is to explain the many different factors which lead to mastitis and poor milk quality. If the farmer, vet or herdsman appreciates the way in which mastitis occurs, then he will be in a much better position to understand and implement the control measures required. Mastitis can never be eradicated. This is because environmental infections such as *Escherichia coli* (*E. coli*) will always be present. It is also highly unlikely that a single all-embracing vaccine will ever be found to suppress the multiplicity of types of infection involved. Control must therefore be based on sound management, and sound management originates best from a thorough understanding of the principles of the disease involved.

The primary objective of this book is the achievement of a thorough understanding of mastitis. If this results in a reduction in the incidence of infection and in so doing benefits both the economics of dairy farming and the welfare of the cow then the authors will be well pleased.

WHAT IS MASTITIS?

Mastitis simply means 'inflammation of the udder'. Most farmers associate mastitis with an inflamed quarter together with a change in the appearance of the milk. These changes are due to the effect of the cow's inflammatory response to infection. However, mastitis can also occur in the **subclinical** form. This means that although infection is present in the udder there are no visible external changes to indicate its presence.

Much of the information needed to reduce the incidence of mastitis has been available for the last 30 years. Research work carried out during the Mastitis Field Experiment trials (MFE) at the National Institute for Research into Dairying (NIRD)* in the 1960s formed the basis of the important mastitis control measures used today, including the proven five point plan which recommended:

1. treating and recording all clinical cases.
2. dipping teats in disinfectant after every milking.
3. dry cow therapy at the end of lactation.
4. culling chronic mastitis cases.
5. regular milking machine maintenance.

Over the past 25 years, great progress has been made in mastitis control in the United Kingdom, mainly due to the uptake of the five point plan by dairy farmers. The clinical incidence has decreased from 121 cases per 100 cows per year in 1968, to about 50 in 1995. One case is one quarter affected once.

There are two basic types of mastitis: contagious and environmental. The greatest progress has been in reducing

* Now called the Animal Grassland Research Institute

the incidence of **contagious** mastitis. Cell counts (also referred to as somatic cell counts and SCCs) relate to the level of contagious infection and so the effect of this progress can be seen in the decrease in the national average cell count in England and Wales from 571,000 in 1971 to 260,000 in 1994. This is shown in Figure 1.1.

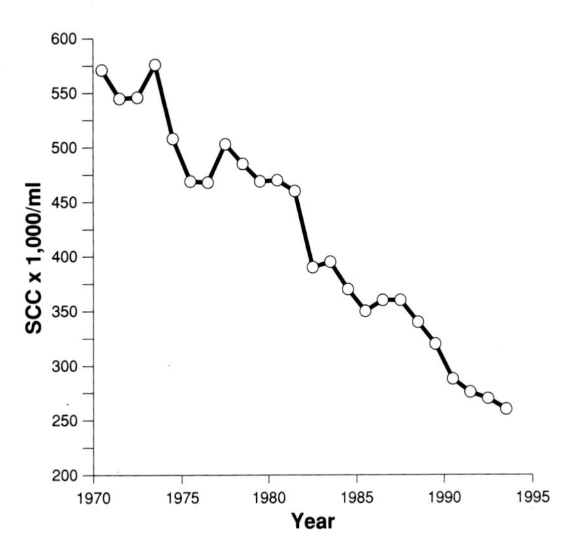

*FIGURE 1.1 Average annual cell count (SCC) for England and Wales 1971 to 1994. (1)**

The aim now is to further reduce contagious mastitis and cell counts, and also to reduce environmental infections. The incidence of **environmental** mastitis has remained unchanged since 1960. This is largely due to an increase in herd size and higher milk yields. Milk yield is correlated to the speed of milking and flowrates have doubled over the past 40 years. Over the same period, faster milking speeds have lead to a 12-fold increase in mastitis susceptibility.

It is therefore a credit to farmers that they have improved the cow's environment and hygiene sufficiently to have prevented an increase in the clinical mastitis incidence over this period. As yields are likely to increase further in the future, the risk of new infections will continue to rise.

Cell counts in different countries can be compared from data collected by the International Dairy Federation (IDF), which is shown in Figure 1.2. The cell count is much lower in countries where a financial incentive is paid to farmers producing quality milk.

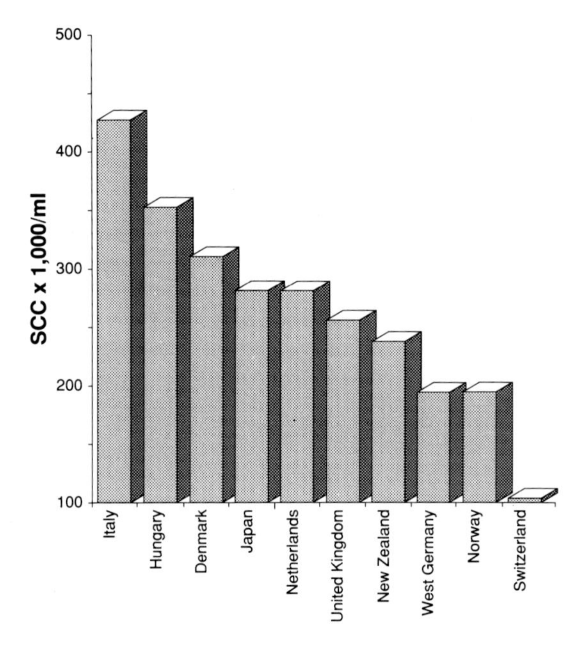

FIGURE 1.2 A selection of national average cell counts, 1993. (2)

Mastitis leads to a reduction in the useful components of milk and increases the level of undesirable elements. This is of course exactly the opposite of what the dairy farmer is trying to achieve. Overall, mastitis results in a less acceptable product and so the value of this milk is much reduced.

Table 1.1 shows the effect of clinical and subclinical mastitis (i.e. raised cell count) on various milk components. It indicates that the yield of lactose and butterfat is reduced substantially. While the total protein level remains little

* Bracketed numbers refer to references which are found on pages 185 and 186.

changed, the level of casein is decreased by up to 18%. This is of great significance to dairy manufacturers, especially cheese makers, as it reduces the manufacturing yield from milk. The changes in butterfat and lactose levels are of great economic significance to the farmer as they make up the basis of his milk price. Mastitis may cause a reduction in butterfat and protein, lowering the price of milk by up to 15%. This will have quite an effect on profit.

Mastitis also produces increased levels of the enzymes lipase and plasmin which break down milk fat and casein respectively and therefore have a significant effect on manufacturing yield and keeping quality. These elements are of utmost concern to milk buyers and in the future it is possible that milk will be tested for plasmin and lipase and producers penalised for high levels of these enzymes.

ECONOMICS OF MASTITIS

Mastitis affects the farmer economically in two ways: through direct costs and indirect costs.

Direct costs:

1. discarded milk.
2. drug and veterinary costs.

Indirect costs:

1. decreased milk yield during remainder of lactation due to udder damage and/or subclinical infection.
2. penalties because of increased cell count.
3. extra labour requirements for treating and nursing.
4. higher culling and replacement rates leading to loss of genetic potential.
5. deaths.

The costs of a clinical case of mastitis have been quantified: in 1994 it was estimated that the average cost of one case of mastitis was between £60 (ADAS) and £168 (Esslemont) (*3*). An average cost of £90 is a well-accepted figure for 1995.

This work assumed that there were three categories of mastitis: mild, severe and fatal, with an incidence of 70%, 29% and 1% respectively. The most common form of mastitis is the mild case which responds quickly to farmer treatment. The costs here include intramammary tubes, discarded milk and a reduced yield for the remainder of the lactation. A severe case of mastitis requires veterinary treatment, while a fatal case of mastitis not only requires veterinary treatment but the cow never returns to the milking herd as she either dies or has to be culled.

Table 1.1 The effect of mastitis on milk components (*4*)

	Components	*Effect of subclinical mastitis*
Desirable	Total proteins	Decreased slightly
	Casein	Decreased between 6 and 18%
	Lactose	Decreased between 5 and 20%
	Solids not fat (SNF)	Decreased by up to 8%
	Butter fat	Decreased between 4 and 12%
	Calcium	Decreased
	Phosphorous	Decreased
	Potassium	Decreased
	Stability and keeping quality	Decreased
	Taste	Deteriorates and becomes bitter
	Yoghurt starter cultures	Inhibited
Undesirable	Plasmin (degrades casein)	Increased
	Lipase (breaks down fat)	Increased
	Immunoglobulins	Increased
	Sodium	Increased – hence the 'bitter' taste

In addition to the cost of mastitis, there are extra risk factors that should be considered. These include high total bacterial counts (TBCs) and the risk of antibiotic residues entering the bulk milk supply. Both of these incur financial penalties.

The majority of the losses in high cell count herds are from subclinical infection resulting in depressed production and reduced yields of lactose, casein and butterfat. It is generally accepted that herds with a cell count of 200,000 or less will have no significant production losses due to subclinical infection. For every 100,000 increase in cell count above 200,000, there will be a reduction in yield of 2.5% (see Figure 9.2, page 120). This reduction together with financial penalties imposed for elevated cell counts can be quite substantial. Some producers in the United Kingdom are now being penalised for supplying milk with a cell count above 100,000.

The average incidence of clinical mastitis in the United Kingdom in 1995 was about 50 cases per 100 cows per year, ranging from some herds with virtually no mastitis to others with up to 250 cases per 100 cows per year.

WHAT ARE REALISTIC PRODUCTION TARGETS FOR THE FUTURE?

The consumer and the dairy companies are requiring milk of increasing quality. In the future it is likely that the dairy companies will continue to lower the production thresholds for TBCs and cell counts, above which producers will incur financial penalties. The importance of cell counts to the dairy companies can be seen from the large financial penalties that they are imposing. Milk Marque, which is the largest milk purchaser in the United Kingdom, has imposed a scale of penalties that will indicate to any farmer that they are only interested in quality milk. Their scale of penalties announced in January 1995 is shown in Table 1.2. There will be an escalating scale of penalties imposed for producers with high cell count milk. Dairy farmers with counts over 500,000 from April 1996 will face penalties equivalent to nearly 25% of the milk price.

The farmer is therefore encouraged to keep reducing the cell count and TBC, and in so doing will ensure that he receives the premium price for his milk. This also benefits the consumer and the dairy industry who will have a quality product with a good shelf life, suitable for manufacturing.

With good herd management it should be possible to have an incidence of clinical mastitis below 30 cases per 100 cows per year, a herd cell count of under 200,000 and TBCs under 10,000. For 'problem' herds this may take several years to achieve. Meeting these goals will improve profitability while ensuring a healthy future both for the dairy farmer and his cows.

Table 1.2 The scale of cell count penalties for Milk Marque in pence per litre (ppl)

Cell count × 1,000/ml	*Old system*	*April to September 1995*	*October 1995 to March 1996*	*April 1996*
under 250	nil	+ 0.2	+ 0.2	nil
251–400	nil	nil	nil	–0.2
401–500	–0.5	–0.5	–1.0	–2.0
501–1,000	–1.0	–3.0	–4.0	–6.0
over 1,000	–2.0	–3.0	–4.0	–6.0

CHAPTER TWO

Structure of Teats and Udder and Mechanisms of Milk Synthesis

STRUCTURE OF THE TEATS AND UDDER

The udder (or mammary gland) is derived from a highly modified sweat gland. As such, the inside lining of the teats and ducts of the mammary gland are essentially modified skin.

Milk is produced by cells lining the alveoli – small sac-like structures deep within the mammary gland (Figure 2.1). Surrounding the alveoli are myoepithelial or muscle cells (see also Figure 2.9). When the stimulus for milk let-down occurs, these cells contract, squeezing milk from the alveoli into the ducts. From there milk flows into the gland (or udder) cistern and then into the teat itself, where it is ready to be drawn from the udder. In higher yielding cows particularly, there will, of course, be some milk stored in the ducts, cisterns and teats between milkings. The mechanisms of milk synthesis are described on page 13.

DEVELOPMENT OF THE UDDER

The development of the udder from the birth of the calf to the start of its first lactation can be divided into four phases.

First Isometric Phase

In the young calf the growth and development of the udder proceeds at the same rate as the rest of the body.

First Allometric Phase

This is a sudden increase in the growth of the udder, which as a result begins to develop more rapidly than the rest of the body. This phase occurs at approximately four to eight months old, i.e. around puberty, and is particularly associated with peaks of oestrogen occurring each time the heifer comes on heat. Development at this stage is primarily of the ducts, which lengthen and penetrate the pad of fat which occupies the site of the udder in the prepubertal calf.

Overfeeding at this stage, and prior to it, leads to an excessive pad of fat being laid down at the site of the future udder. This can cause depressed yields later in life. For example, in one trial(5), two groups of heifers were reared to produce liveweight gains of 1.1 kg per day (high) and 0.74 kg per day (conventional). Not only did the mammary glands of the conventionally reared heifers weigh more (40% more) but they also contained much more secretory tissue (68% more). Gross overfeeding of young heifers is therefore to be avoided. It is thought that a diet high in forage during rearing stimulates greater rumen development and higher appetite capacity at maturity. Protein intakes should be high (for example, 18% crude protein) and of good quality to promote udder development.

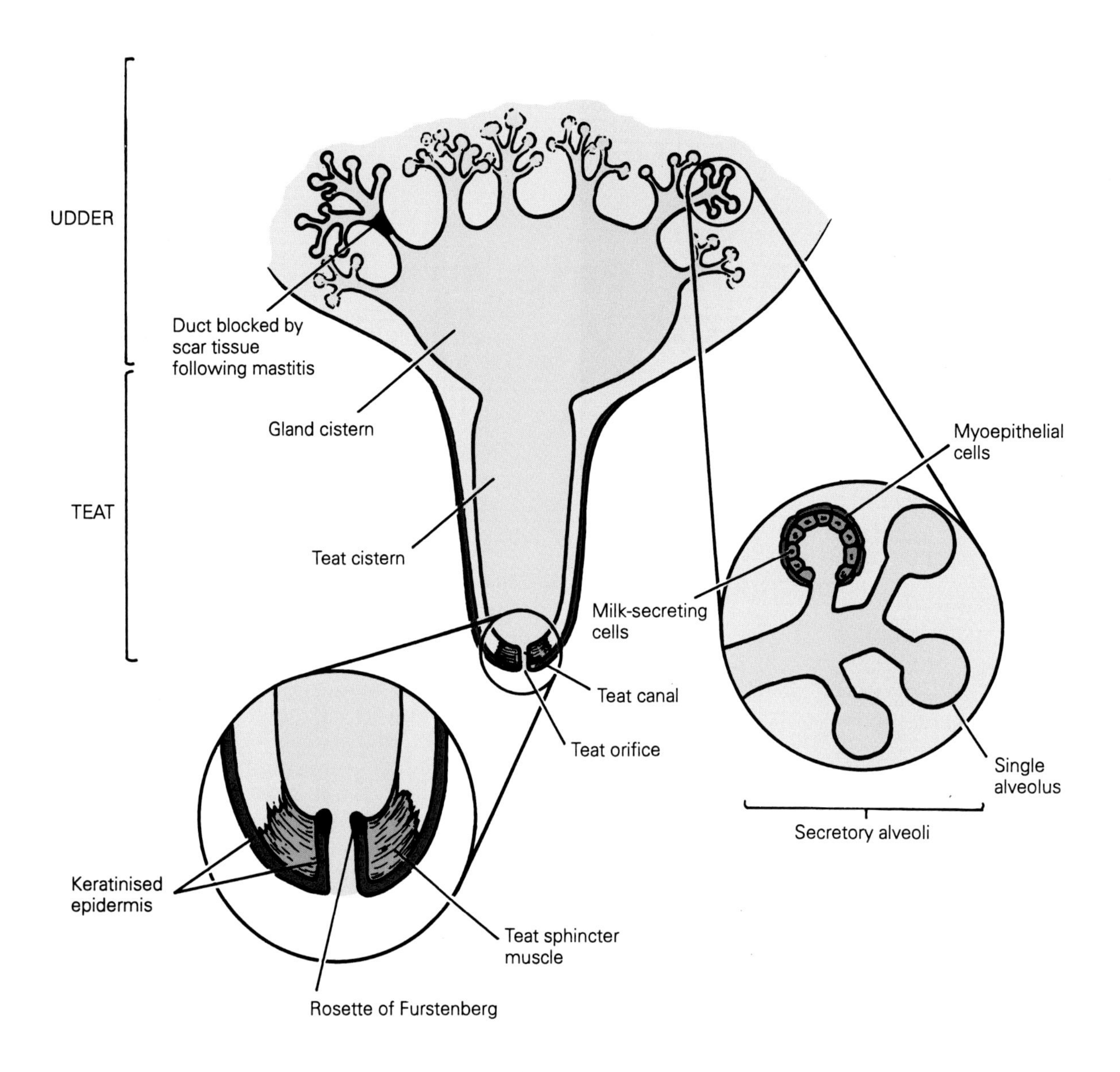

FIGURE 2.1 The structure of the udder and teat.

Second Isometric Phase

From after the onset of puberty until the beginning of pregnancy the udder grows at the same rate as the other body organs.

Second Allometric Phase

Following conception, udder development once again becomes rapid, with the highest growth rate occurring from mid-pregnancy onwards. During this phase the cells of the alveoli especially become more developed and change into a tissue type which is able to secrete milk.

SUSPENSION OF THE UDDER

The udder consists of four separate mammary glands, each with its own distinct teat. There is **no** flow of milk from one quarter to another, neither is there any significant direct blood flow from one quarter to another. The blood supply to the udder is massive, with some 400 litres of blood flowing through the udder to produce each litre of milk. Add to this the weight of the secretory tissue and of milk stored and it is easy to see how udder weights of 50 to 75 kg are obtained.

The reason why all milk should be discarded when treating one quarter with antibiotics is that antibiotics may be absorbed from that quarter into the bloodstream, travel around the body and then be deposited back into one of the other untreated quarters. The amount of antibiotic involved is, of course, relatively small, but it may be enough to lead to bulk tank failure.

The suspension of the udder is very important. It is shown in Figure 2.2 and consists of:

- the skin, which only plays a very minor role.
- superficial lateral ligaments. These originate from the bony floor of the pelvis and pass down the outside of the udder, especially at the front and the sides. They branch out attaching to the abdomen (in front) and the inner thighs (from the sides).
- deep lateral ligaments which also originate from the tendon around the floor of the pelvis. Passing down the outside of the udder (but inside the superficial ligaments) they send small 'cups' across within the gland and these eventually connect to similar branches from the central median ligament. The largest branch sweeps under the base of the udder, just above the teats, to join the median ligament and provides the major udder suspension.
- the median ligaments which attach to both the pelvic floor and the abdominal wall muscles and pass down the centre of the udder. At the base they separate and join the lateral ligaments at the right and left sides of the udder. Branches also connect to connective tissue which separates the fore and hind quarters. The median ligaments contain elastic fibres which allow a degree of 'give', providing a shock-absorber effect and allowing the udder to expand as milk accumulates between milkings.

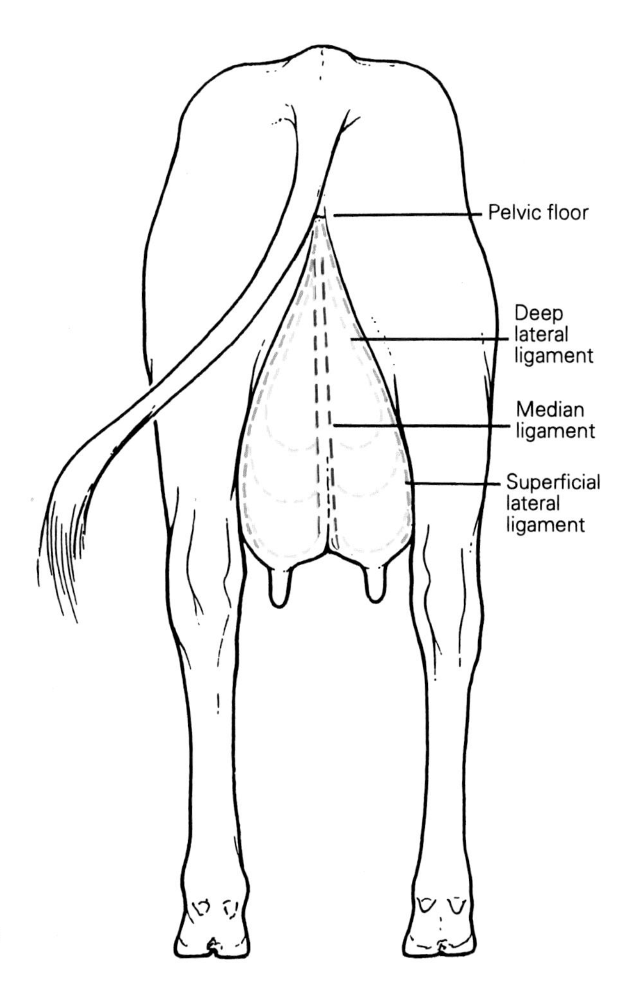

FIGURE 2.2 The suspension of the udder.

Rupture of the Suspensory Apparatus

Rupture of the ligaments may occur gradually or spontaneously and may be associated with a variety of factors, the most important of which are:

- age: the elastic tissue in the median ligaments especially, deteriorates with age.
- over-engorgement and oedema of the udder (see pages 170–172 for the many causes of udder oedema). This is one good reason why heifers and cows should not be 'steamed up' (fed extra concentrates) excessively or be kept overfat before calving.
- poor conformation: it is important to select for a 'type' which has good udder attachment and evenly placed front and rear teats.

Rupture of the median ligaments is probably the most common reason for poor udder suspension. It leads to loss of the 'cleavage' between quarters causing the teats to splay outwards (see Figure 2.3 and Plate 2.1), making it difficult to attach the milking units. It also often leads to air leakage during milking, especially when the unit is first applied, thus producing teat-end impacts (see page 63) and increasing the mastitis risk.

Rupture of the deep lateral ligaments is invariably associated with concurrent rupture of the superficial ligaments and leads to a total drop of the whole udder (see Figure 2.4 and Plate 2.2). The teats drop to well below hock level and can easily become injured as the cow walks.

Rupture of the anterior ligaments (the front portions of the superficial and deep ligaments) occurs less frequently. It is seen as a gross enlargement at the front of the udder (see Figure 2.5 and Plate 2.3) which often (but not always) leads to a dropping of the front teats. Conditions such as haematomas (which are large accumulations of blood under the skin) and rupture of the abdominal wall can sometimes be confused with rupture of the anterior udder ligament.

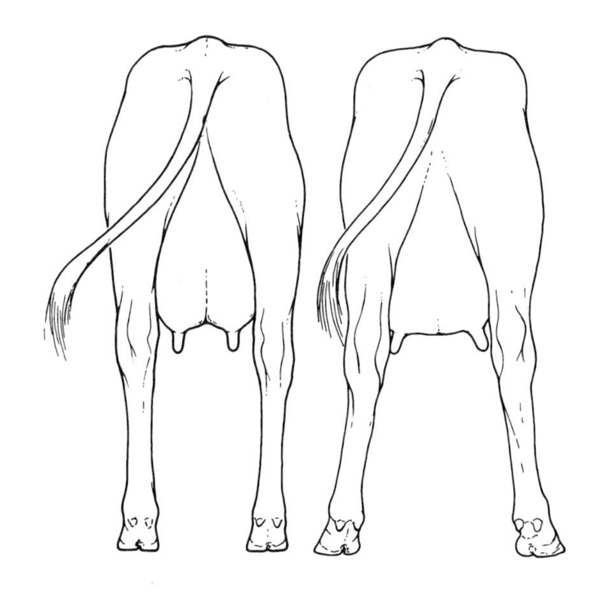

FIGURE 2.3 Rupture of the median suspensory ligament (right) leads to splaying of the teats and loss of normal udder cleavage as shown on the left.

PLATE 2.1 Rupture of the median udder ligament, leading to splaying of the teats.

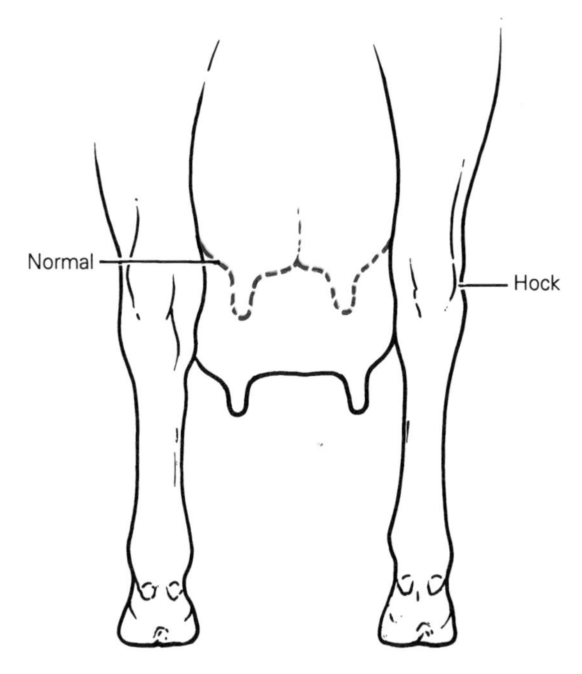

FIGURE 2.4 Rupture of the deep lateral ligaments leads to the udder dropping to well below hock level.

PLATE 2.2 Rupture of the median and lateral ligaments, leading to a total 'drop' of the udder.

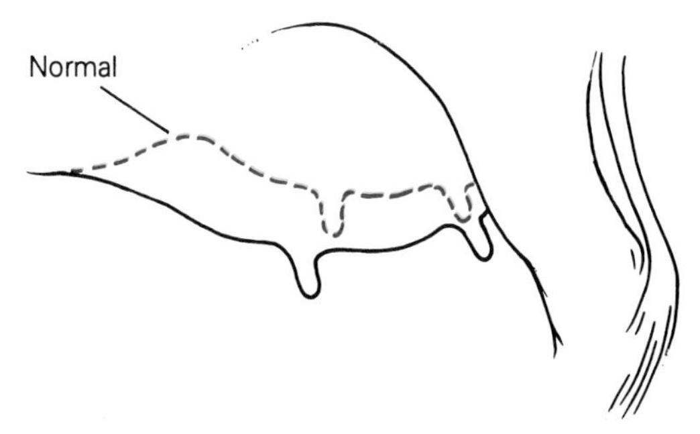

FIGURE 2.5 Rupture of the anterior portion of the deep lateral ligament is seen less commonly than that of the other ligaments. It leads to a swelling in the front of the udder and the front teats drop.

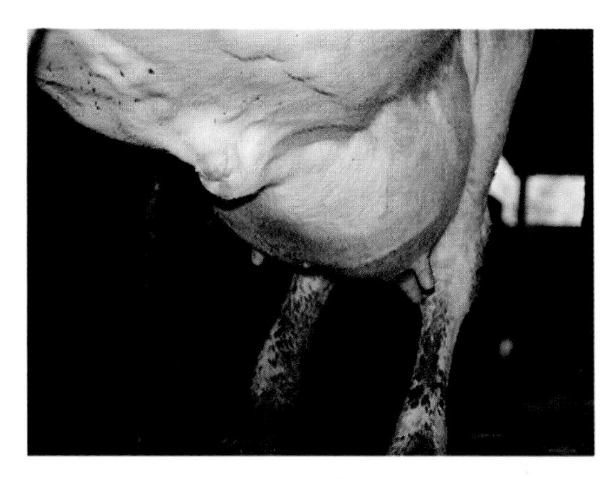

PLATE 2.3 Rupture of the anterior ligament – a large swelling appears at the front of the udder.

THE TEATS

The cow has four main teats, with 60% of production coming from the two hind teats. There may be varying numbers of supernumerary teats (extra teats).

Supernumerary Teats

These are congenital (ie the calf is born with them) and often inherited and hence it is advisable not to select from cows with large numbers of supernumerary (also known as accessory) teats.

Supernumeraries should be removed at the same time as the calf is disbudded. The calf needs to be sitting upright in order to allow a thorough inspection of the udder: if simply looked at from between the hind legs while standing it is very easy to miss those accessory teats which are situated between the main teats. If in any doubt over which are the supernumeraries, simply roll the teat between your finger and thumb. The supernumerary is much thicker and has no palpable teat cistern. They are most easily removed by lifting the skin under their base with your finger and simply cutting them off using curved scissors. No anaesthetic is required in animals less than two months old.

Failure to remove accessory teats has several disadvantages:

- they are unsightly and affected animals are less saleable.
- they may develop mastitis (especially summer mastitis – see Chapter 13) and cause an abscess on the udder.
- if situated very near to, or at the base of a true teat, they may interfere with milking and lead to air leakage with resulting teat-end impacts (see page 63).

Functions of the Teat

The most important function of the teat is to convey milk to the young calf. Its use has of course been modified, to allow hand and machine milking to produce food for man.

The teat has an erectile venous plexus at its base (Figure 2.6). This is a mass of interconnecting blood vessels which, under the stimulus of suckling and milk let-down, becomes engorged to produce a more rigid and turgid teat. The stiff teat is extremely important in both suckling and machine milking. If the teat were to collapse it would impede the flow of milk from the gland cistern into the teat cistern and hence slow down the milking process. Many herdsmen have probably seen how much blood the erectile venous plexus can hold: a cow with a cut at the tip of the teat bleeds very little, whereas a cut through the venous plexus at the base bleeds profusely and can occasionally lead to serious, or even fatal, blood loss.

Secondly, and as we shall see in later sections, the teat and especially the teat canal, have very important functions in preventing the entry of infection into the udder.

Finally, the teat is richly innervated and hence can rapidly convey suckling stimuli to the brain, thus inducing good milk let-down. This rich innervation can occasionally make handling cows with highly sensitive cut teats somewhat hazardous!

Teat Size

As one would expect, this varies enormously with lengths ranging from 3 to 14 cm. The diameter also varies from 2 to 4 cm. Teat length increases from the first to the third lactation and then remains constant. On both small, short teats and long, wide teats it may be difficult to get good liner attachment and hence there is an increased risk of liner slippage and teat-end impacts.

Teats may be cone-shaped and pointed or cylindrical with a flat tip (Figure 2.7). Cylindrical teats are said to be less prone to mastitis and are certainly the most common.

The Teat Wall

The teat wall consists of four layers, each having an important function in milk let-down and mastitis control.

The epidermis

This is the thick outer lining of the skin. Its surface consists of a layer of dead, keratinised cells (Figure 2.6) which produce a hostile environment for bacterial growth. (Keratin is a sulphur-containing protein which impregnates cells, thereby increasing their strength. It is also present in hair, horn and hoof.)

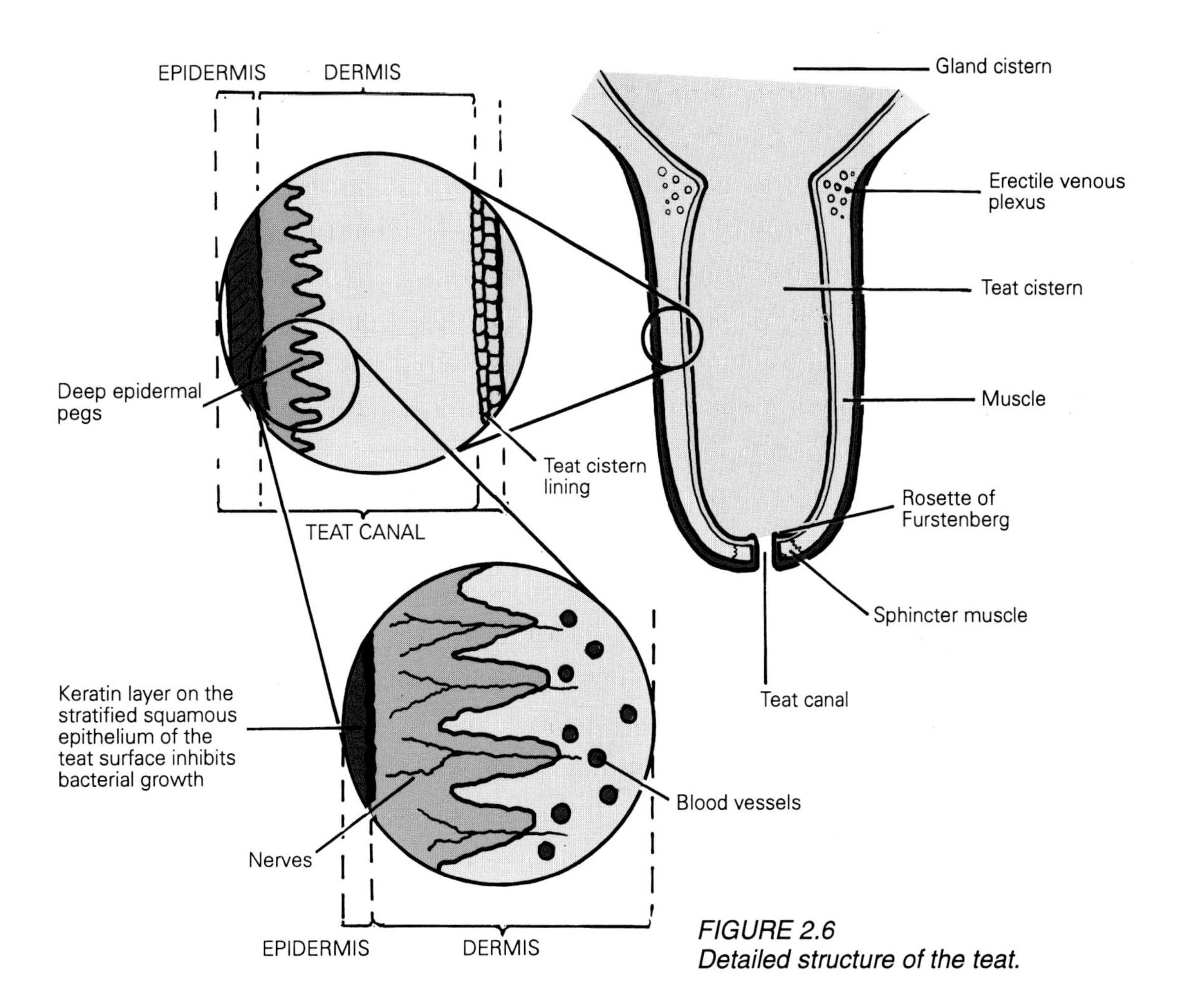

FIGURE 2.6
Detailed structure of the teat.

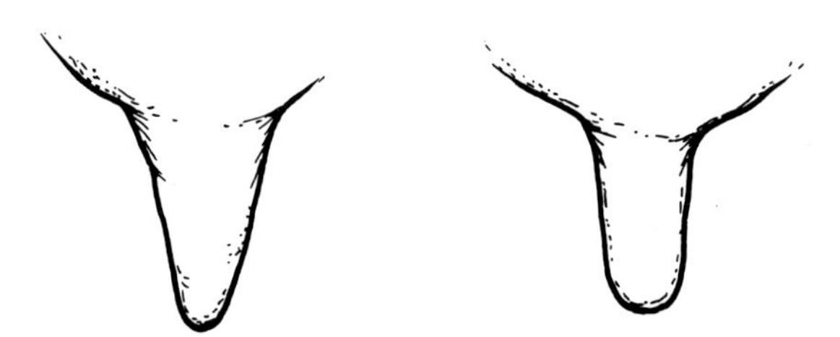

FIGURE 2.7 Teats may be cone-shaped (left) or cylindrical. Cylindrical teats are said to be less prone to mastitis.

All skin is lined with a keratinised epidermis, but teat skin has a particularly thick layer, some four or five times thicker than that of normal skin. It is also very firmly anchored to the underlying dermis (or second layer of skin) by deep epidermal pegs, or papillae. If the skin of the udder is pinched between the finger and thumb, it moves very freely over the underlying tissue. Try doing the same with teat skin: it is firmly attached. The epidermis of the lips and muzzle of the cow has a similar structure. It is thought that the firm attachment of the epidermis protects the teat from the sheer forces involved in suckling and machine milking and also reduces the chances of injury due to physical trauma. Even so, it is surprising how frequently teats get damaged.

Teat skin has no hair follicles, no sweat glands and no sebaceous glands. In practical terms this means that teat skin is particularly susceptible to drying and cracking, which is one reason why an emollient is necessary in teat dips (see also page 98). It also means that there is little or no flow of sebum over teat skin and hence fly repellents should be applied directly onto the teats. Ear tags and pour-on preparations give a very poor flow of insecticide onto the teats.

The dermis

This is the second layer of the teat wall and is the tissue which carries the blood vessels and nerves. However, the fine sensory nerve endings are in the epidermis, which is why exposure of an eroded epidermis (for example a teat sore) can be so painful. At the base of the teat, adjacent to the udder, the dermis contains the erectile venous plexus.

The muscles

There is a variety of muscles which are set in transverse, oblique and longitudinal planes in the dermis of the teat wall. There is also a circular sphincter muscle around the teat canal. During milking, the teat elongates and the canal opens but becomes shorter. After milking, muscle contraction leads to a shortening of the overall teat and closure of the teat sphincter but a lengthening of the teat canal. The shortened teats are less prone to physical trauma and the lengthened and closed canal reduces the risk of entry of bacteria. These changes are shown in Figure 2.8.

Teat cistern lining

The teat cistern is lined with cuboidal epithelium, that is a double layer of 'block' cells (Figure 2.6). In the normal cow these are held tightly together; however in response to bacterial invasion they have the ability to move slightly apart, which allows the entry of infection-fighting white blood cells from the small blood vessels beneath (see page 23).

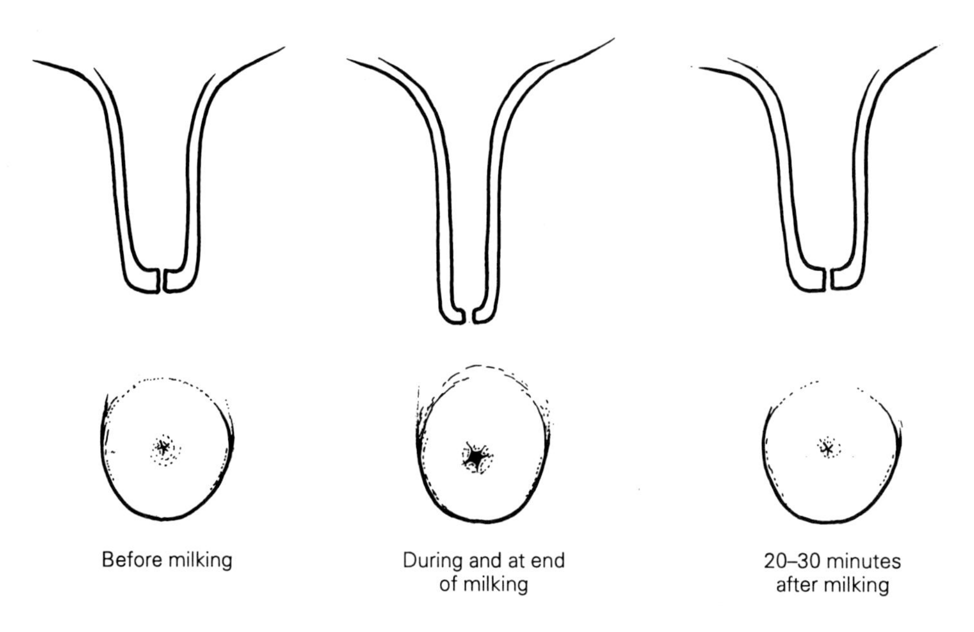

FIGURE 2.8 Changes in teat length and shape during milking. At the end of milking the interdigitating folds of the teat canal close tightly.

The Teat Canal

Sometimes also known as the streak canal, the teat canal is approximately 9 mm long (range 5–13 mm). Shorter canals have an increased mastitis risk. The canal is lined with the same keratinised epidermis (layers of flattened plate-like cells) as the teat skin. This is arranged in a series of folds which interdigitate to produce a seal when the muscle around the teat sphincter constricts at the end of milking (Figures 2.1 and 2.8). Elastic fibres at the teat end assist in the closure of the canal.

At the internal entrance to the teat canal is the Rosette of Furstenberg (see Figure 2.1). This is a ring of tissue containing lymphocyte and plasma cells which are important in the recognition of foreign invading bacteria and stimulate the first stages of an immune response.

MILK SYNTHESIS AND HOW IT IS AFFECTED BY MASTITIS

This is an enormously complex process and will not be covered in detail.

Milk is synthesised in the cells lining the alveoli, which are the small sacs at the very end of the ducts deep within the udder (see Figures 2.1 and 2.9). The average composition of milk is shown in Table 2.1.

Table 2.1 Approximate composition of milk from Friesian/Holstein cows

	Milk
Total solids	12.5%
Protein	3.3%
Casein	2.9%
Lactose	4.8%
Ash	0.7%
Calcium	0.12%
Phosphorus	0.09%
Immunoglobulins	1.0%
Vitamin A μg/g fat	8
Vitamin D μg/g fat	15
Vitamin E μg/g fat	20
Water	87.5%

Lactose

Glucose is produced in the liver, primarily from propionate, a product of rumen fermentation. After it is transferred to the udder, part of the glucose is converted into another simple sugar, galactose. Next, one molecule of glucose combines with one of galactose to produce lactose. This is then known as a disaccharide (viz. two sugars conjoined).

- liver – propionate to glucose
- udder – glucose to galactose
- udder – glucose + galactose = lactose

Lactose is the main osmotic determinant of milk (the factor governing the concentration of its components in solution). To maintain milk at the same concentration as blood, lactose increases and decreases as the strength of the other milk components vary. However, the pH of milk is slightly lower than blood (i.e. more acidic):

- pH of blood = 7.4
- pH of milk = 6.7

This difference may be used to attract drugs such as erythromycin, trimethoprim and penethamate into the mammary gland as lower pH solutions are drawn to those with higher pHs.

If lactose concentrations in the udder fall (as occurs with mastitis) then sodium and chloride levels increase to maintain the osmotic pressure of the milk. This is one of the causes of the bitter and slightly salty taste of mastitic milk. (Some farmers occasionally taste the milk of cows they are intending to purchase, in an attempt to identify mastitis). These changes can also be used to help assess mastitis status by electrical conductivity measurements, since sodium and chloride are much better conductors of electricity than lactose.

Protein

The majority of protein in milk is in the form of **casein**, which is synthesised in the udder cells before transfer into the ducts. However, it is the *energy* content of the diet which has the major effect on the casein content of milk. Dietary protein has relatively little influence on milk protein content. Other types of protein present in milk in small quantities are albumin and globulins. These are transferred directly from blood into milk.

Mastitic milk has a reduced casein content but increased levels of albumin and globulin. The total protein content of the milk may remain constant therefore, but the milk is of much poorer quality, particularly for manufacture. This is because the coagulation of casein is so important as a starting process for cheese and yoghurt production. In addition, mastitic milk contains increased levels of the enzyme **plasmin**, which decomposes casein in stored milk. Unfortunately plasmin is not destroyed by pasteurisation and it remains active even at 4°C (storage temperature in supermarkets). Mastitic milk will therefore continue to be degraded even following pasteurisation and storage at 4°C: this explains why manufacturers are prepared to pay a premium for low cell count milk.

Fat

Milk fat is formed in the udder secretory cells when fatty acids are combined with glycerol and converted into a neutral form of fat called triglyceride.

Glycerol + 3 fatty acids = triglyceride

Fatty acids are derived from three main sources:

- body fat (50% of total fatty acids). Hence body condition score is an important determinant of milk fat levels, especially in early lactation.
- dietary fat, especially long chain fatty acids (those which are solid at room temperature and are components of butter and lard). The use of protected fat (viz. fat which has been treated so that it can pass through the rumen unchanged) can therefore increase the butterfat content of milk.
- finally, fatty acids are synthesised in the udder from acetate which is absorbed as a product of rumen fermentation. High fibre diets, which promote increased levels of acetate in the rumen, will therefore lead to an increase in milk fat production.

Small particles of milk fat (triglyceride) are extruded from the secretory cells in the alveoli and are covered by a thin protein membrane before passing into the milk (Figure 2.9).

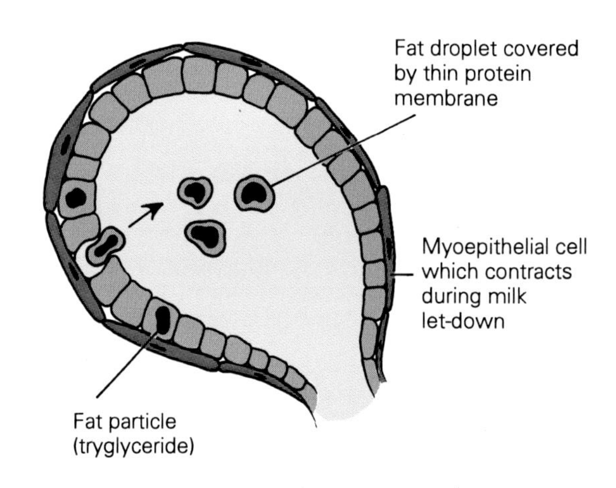

FIGURE 2.9 The synthesis of milk fat droplets in the alveolus.

Mastitic milk has an increased level of the enzyme **lipase**. This leads to degradation of the milk fat into its fatty acid components and thus imparts a rancid flavour to the milk. Increased levels of fatty acids can inhibit starter cultures used in cheese and yoghurt manufacture and can also impart a rancid flavour to these products.

Minerals

The minerals in milk are derived directly from the blood. Calcium is actively secreted in association with casein.

CONTROL OF MILK SYNTHESIS

In most mammals initiation of lactation and continued milk production is controlled by the hormone **prolactin**. In the cow, however, continued milk secretion is influenced by a complex interaction of steroids, thyroid hormone and growth hormones, the latter being more commonly known as bovine somatotrophin (BST). BST is a natural hormone synthesised by the pituitary gland, a small organ at the base of the brain. Higher yielding cows have more BST circulating in their blood than lower yielding cows, and cows at peak yield more than late lactation animals. BST can now be produced synthetically and at the dose rate currently being suggested increases yields by 10–20%, i.e. 4–6 litres per day. BST alters the cow's metabolism, so that a greater proportion of her food is used for milk production, thus making her more efficient. Some 4–6 weeks after starting dosing and after an initial increase in yield, there is an increase in food intake and appetite.

Milking Frequency

Increased frequency of milking also increases yield. Changing from twice to three times daily will increase production by around 10–15% in cows and 15–20% in heifers. Because of the flatter lactation curve it produces, three times a day milking has to be continued to the end of lactation to obtain its full beneficial effect.

Reducing milking frequency decreases yield. For example, if cows are only milked once a day, yields may fall by up to 40%. The majority of farms milk at intervals of 10 hours and 14 hours. Trials have suggested that this does not produce significantly lower production than precise 12-hourly milking in anything but the highest yielding cows.

The influence of milking frequency on milk yield appears to be controlled by local mechanisms acting within the udder. This is known to be true because if two quarters are milked twice daily and the other two are milked four times daily, only the four times daily quarters show an increase in yield. Initially it was thought that back-pressure of milk within the alveoli was responsible. However, if the milk withdrawn from the four times daily quarters is replaced by an equal volume of saline (i.e. to restore the pressure within the alveoli), yields still increase. It has now been shown that milk naturally contains an **inhibitor protein** and it is the presence of this inhibitor, acting directly on the secretory cells within the alveoli, which influences yield. More frequent milking leads to more frequent removal of the inhibitor protein and hence more milk is produced. Not only does frequent removal of the inhibitor protein stimulate increased activity of secretory tissue (and hence increased yields), but it also slowly increases the **amount** of secretory tissue present, producing a longer term effect.

The extent of these effects depends partly on the internal anatomy of the udder. An udder with large teat and gland cisterns and large ducts, will store less milk in the alveoli between milkings. There is then less contact between milk inhibitor protein and the secretory tissue and hence the cow will be a higher yielding animal. In the average cow approximately 60% of the total milk is stored in the alveoli and small ducts and 40% in the cisterns and large ducts.

Although not yet feasible, vaccination of cows against their own inhibitor protein raises interesting possibilities, as this could be a further way of increasing yields.

Environmental Temperature

Under very cold conditions, water consumption and therefore milk yield falls. When the weather is very hot, food (and especially forage intake) falls and this can depress both milk yield and milk fat levels. High environmental humidity exacerbates the effects of both hot and cold weather.

Length of Dry Period

Towards the end of lactation, the number of active alveolar secretory cells slowly decline, reaching a minimum during the early dry period. The alveolar cells do not die, but simply collapse, so that the space within the alveolus disappears and the udder consists of a greater proportion of connective tissue. New secretory tissue is laid down when the cow starts to 'freshen' ready for the next calving and hence the total amount of secretory tissue (and therefore yield) increases from one lactation to the next. A dry period of 6–8 weeks is ideal. If the cow is not dried off at all, the next lactation yield may be as much as 25–30% lower. This may occur, for example, following an abortion, or if a bull is running with the herd and no pregnancy diagnosis (PD) is carried out.

CHAPTER THREE

Teat and Udder Defences Against Mastitis

Considering how often the teats, and especially the teat ends, become contaminated with bacteria, it is surprising that mastitis infections are not more common. This chapter studies the many ways in which the cow repels infection.

THE TEAT SKIN

Teat skin has a thick covering of stratified squamous epithelium (Figure 2.6), the surface of which consists of dead cells filled with keratin. When intact this provides a hostile environment for bacteria, thus preventing their growth. In addition, there are fatty acids present on skin which are bacteriostatic, that is they prevent bacterial growth. However, these can be removed by continual washing, especially using detergents, and this is one of the reasons why an increasing number of people now 'dry wipe' before milking, rather than wash the teats.

The normally intact surface of the skin may also become compromised by cuts, cracks, chaps, bruising, warts, pox lesions etc. Bacteria can then multiply on the surface of the skin and become a reservoir for mastitis infections. This is particularly the case for organisms such as *Streptococcus dysgalactiae* and *Staphylococcus aureus*. Maintaining an intact and healthy teat skin is one of the important functions of the emollient present in post-milking teat dips.

THE TEAT CANAL

The teat canal is also lined with keratinised epidermis (see Figures 2.1 and 2.6) and so it has similar defence properties to teat skin.

These antibacterial properties of the teat canal are at their most effective when the canal is closed. This is when the action of the sphincter muscle has caused the adjoining folds of stratified squamous epithelium to interlock (see Figure 2.8) and form a seal. After milking, a waxy plug of keratin is secreted into the canal, thereby replacing any residual plug of milk, and further reducing the chances of bacterial penetration of the teat canal between milkings. It takes at least 20–30 minutes for the teat end to become fully closed and hence advice is often given that cows' teats should be kept clean and that animals should not be allowed to lie down until at least 30 minutes after milking.

Figures 3.1a and b show the importance of teat sphincter closure in relation to *Escherichia coli* mastitis. Teats were dipped in a broth culture of *E. coli* at varying times after milking. Of the teats dipped and exposed to *E. coli* in the first 10 minutes after milking, 35% developed mastitis. However, if the teats were not dipped into the *E. coli* broth until a few hours before the next milking, then only 5% developed mastitis.

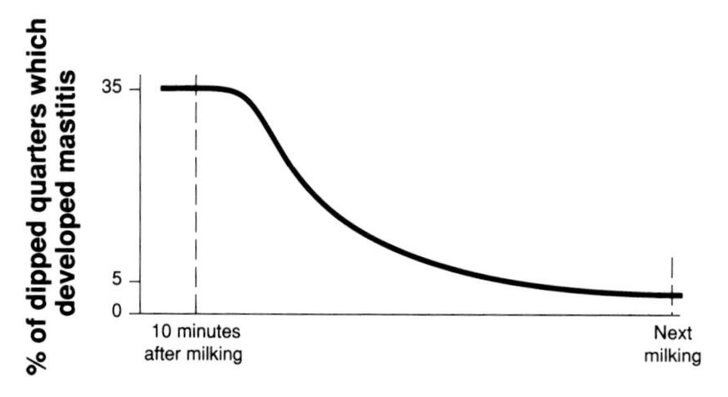

FIGURE 3.1a The importance of teat sphincter closure in relation to E. coli *mastitis: if teats were dipped in a broth culture of* E. coli *0–10 minutes after milking, 35% of quarters developed mastitis. This reduced to 5% if teats were dipped in E. coli broth immediately prior to the next milking. (6).*

This same information can be quantified in terms of the pressure required to force fluid back up through the teat canal and is shown graphically in Figure 3.1b.

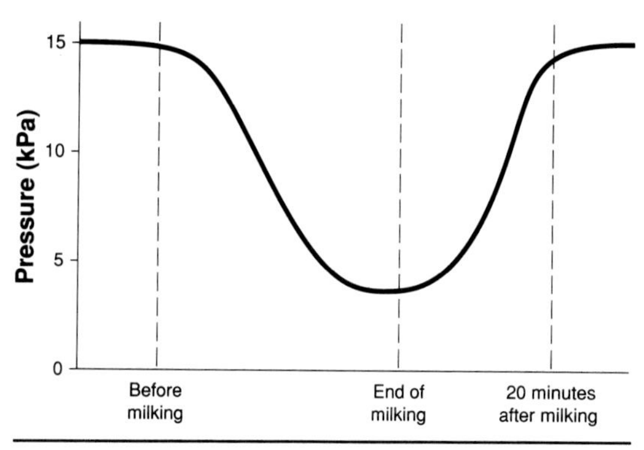

	Pressure required to force bacteria through teat canal (kPa)
Before milking	15
During milking	4–6
20–30 mins after milking	15

FIGURE 3.1b Pressure required to force fluid through the teat canal before, during and after milking.

Because of the ease with which bacteria can penetrate the teat end, it is particularly important to prevent liner slippage and resultant teat-end impacts at the end of milking (see page 63).

Teat Canal Dimensions and Speed of Milking

Cows with short teat canals (i.e. short vertical length) and those with a wide cross section diameter are more susceptible to mastitis. Cows with 'open' teat canals also milk faster. As this is likely to be an inherited feature, there will be a genetic susceptibility to mastitis. Conversely, provided that teat-end lesions do not develop, 'hard' milkers, with slow milk flow rates, will have a lower infection rate than fast milkers.

However, speed of milking is correlated to yield (the greater the yield the greater the milk flow rate) and hence increasing selection for yield has led to an overall increase in milk flow rates. Table 3.1 shows that the average milk flow rate for a fast milker doubled between 1950 and 1990.

Table 3.1 Average milk flow rates of fast milkers (kg/min) (*7*)

	per quarter	*per cow*
1950	0.8	3.2
1990	1.6	6.4

This has led to a **12-fold** increase in mastitis risk over the same period. Yields will undoubtedly increase even further in the future and hence we can expect to see corresponding increases in mastitis susceptibility. It will be a challenge to us all to provide optimum conditions of housing, machine function and management to control mastitis.

Table 3.2 shows numerically the relationship between flow rate and mastitis incidence when teats were experimentally exposed to a high bacterial challenge. Three different

Table 3.2 The influence of milk flow rate from the teat end on the percentage of quarters becoming infected following experimental challenge. A poorly functioning machine dramatically increases the infection rate. (*7*)

Quarter flow rate (kg/min):	< 0.8	0.8–1.2	1.2–1.6	>1.6
Milking conditions				
Good pulsation + shields	3%	4%	7%	36%
No pulsation	15%	20%	38%	92%
Pulsation + impacts	36%	37%	55%	100%

machine conditions and varying flow rates were used. Initial results were obtained with a well-functioning machine where the liners were fitted with teat shields (see Figure 5.7). If there was no pulsation or, even worse, if teat-end impacts were a problem, then the mastitis risk (expressed as the percentage of quarters becoming infected) became greater, reaching 100% in cows with very high milk flow rates. (Milking machine function is discussed in Chapter five.)

Cows with high flow rates are also much more susceptible to contracting new infections during the dry period.

MASTITIS AND MILKING FREQUENCY

The flushing action of milking, namely milk flowing through the teat canal, removes bacteria. This is part of the natural defence mechanism which helps to decrease the risk of mastitis. This is particularly true for adhesive organisms such as *Streptococcus agalactiae* and *Staphylococcus aureus*, for which one mechanism of invasion is thought to be slow growth through the teat canal. Hence cows milked three times daily are generally less susceptible to mastitis than cows milked twice daily.

Increased frequency of milking also decreases the pressure of milk within the udder, especially immediately before the next milking. Excess pressure within the teat shortens the teat canal and in so doing further increases susceptibility to mastitis organisms invading the udder.

This all assumes optimum functioning of the milking equipment. If machine function is poor, with defective pulsation and/or teat-end impacts, then increased frequency of milking would lead to an increased risk of mastitis.

TEAT-END DAMAGE AND MASTITIS

The teat canal is obviously of vital importance in the prevention of new cases of mastitis and clearly it follows that any damage to the teat end will compromise the defence mechanisms. Examples of teat-end damage are described in detail in Chapter 14 and include:

- physical trauma: cuts, crushing or bruising.
- 'black spot': a lesion, probably traumatic in origin, with secondary infection caused by the bacterium *Fusobacterium necrophorum.*
- milking machine damage: teat-end oedema, haemorrhage, sphincter eversion etc.
- hyperkeratosis (increase in volume of the keratinised skin of the teat sphincter) and eversion, associated with factors other than the milking machine.
- excessive dilation of the canal, for example, when administering intramammary antibiotics, or when

inserting a teat cannula. This can produce cracks in the keratin lining, thereby providing an opportunity for bacterial multiplication and penetration of the canal. The way that teat cannulae are used is particularly critical, since it is often 1–2 days after **withdrawal** of the cannula (especially after it has been in situ for several days) that mastitis occurs. This is presumably because the tightly fitting cannula prevents bacterial entry while it is in position, but after removal the stretched canal has lost its bacterial defences, allowing easy entry of infection. For this reason many recommend infusing a small quantity of antibiotic after each milking, for the first 3–4 days following removal of the cannula.

DEFENCES WITHIN THE UDDER

Even when bacteria have managed to overcome the defence mechanisms of the teat canal and have either grown through it or been forced through by the milking machine, clinical or subclinical mastitis is by no means a certainty. There are several highly efficient systems within the udder which assist in the removal of bacteria and often prevent infections becoming established. These can be categorised as intrinsic defence mechanisms, which are systems continually present in the udder, and inducible systems, which come into operation in response to bacterial invasion.

Intrinsic Defence Mechanisms

Lactoferrin

This is an iron-binding protein. In the dry, non-lactating udder lactoferrin prevents the growth of bacteria by removing the iron from udder secretions. Although the risk of new **infections** with *E. coli* in the dry period may be as much as four times greater than during lactation, the development of clinical disease (i.e. *E. coli* mastitis) in dry cows is a rarity due to the effect of lactoferrin (see Table 3.3).

Table 3.3 Experimental *E. coli* infection in lactating and dry cows (*8*)

	No. of quarters challenged	*No. of quarters developing clinical mastitis*
Lactating cows	16	12[1]
Dry cows	12	2[2]

1. Of the four quarters which did not show clinical mastitis, two had a high cell count and two had subclinical mastitis.
2. Both cases were in cows challenged only a few days prior to calving when the lactoferrin in milk had already fallen to a low level.

The bacteriostatic (inhibition of bacterial growth) effects of lactoferrin are lost during lactation because:

- lactoferrin is present in only low concentrations.
- high citrate levels in milk compete with lactoferrin for iron, producing iron citrate. This can be utilised by the bacteria during their growth processes.

Lactoperoxidase

All milk contains the enzyme lactoperoxidase (LP). In the presence of thiocyanate (SCN) and hydrogen peroxide (H_2O_2), lactoperoxidase can inhibit the growth of some bacteria (gram-positive organisms, see page 43) and kill others (gram-negatives). The level of thiocyanate in milk varies with the diet, being particularly high when brassicas and legumes are fed. Hydrogen peroxide can be produced by bacteria themselves. Gram-negative bacteria produce very little H_2O_2, and so the lactoperoxidase system is probably not important in their control. There is some evidence that gram-positive bacteria such as *Streptococcus uberis* may produce sufficient H_2O_2 for the lactoperoxidase system to be partially effective in their control.

Complement

Complement is the general term for a series of proteins which, when acting together, produce a cascade effect which results in the killing of certain strains of gram-negative bacteria, such as *E. coli*. *E. coli* is one of a number of coliforms which can be grouped into serum-sensitive strains (killed by complement) and serum-resistant strains (not killed). It has been shown that only the latter are likely to produce mastitis. If a serum-sensitive strain of *E. coli* is isolated from a milk sample therefore, it is likely to be a contaminant only, and not a cause of mastitis.

Immunoglobulins (antibodies)

Antibodies are unlikely to have a primary role in mastitis control, since it is well known that colostrum contains very high levels of antibodies and yet freshly calved cows can develop peracute mastitis and frequently **do** get severe mastitis several days after calving. The role of specific antibodies against mastitic bacteria is unclear. Probably their main function is in the opsonisation of bacteria prior to their being engulfed by white blood cells and macrophages. Opsonisation is a process whereby the bacteria become coated with antibody. A portion of an antibody molecule (the Fab arm) attaches to the bacteria, leaving a second arm (the Fc fragment) exposed. White blood cells (PMNs) are activated by the exposed Fc arm and attach to it. Phagocytosis (engulfing) of the bacteria can then proceed much more rapidly.

Cells in milk

There is a variety of different types of cells in normal milk, but by no means all of them can kill bacteria. The total number of cells can be counted and is expressed as the somatic cell count (SCC). Approximate percentages are given in Table 3.4 although there is still some dispute concerning which cell types are present.

Table 3.4 Percentage of cell types in milk and colostrum (*9*)

	Mid-lactation	*Colostrum*
PMNs[1]	3	61
Vacuolated macrophages	65	8
Non-vacuolated macrophages	14	25
Lymphocytes	16	3
Duct cells	2	3

[1] PMNs = polymorphonuclear leucocytes, bacteria killing cells, also referred to as neutrophils

PMNs (polymorphonuclear leucocytes, a type of white blood cell) are the most important bacterial killing cells. However, it is generally agreed that they are present in such low numbers in normal milk as to be ineffective against a heavy bacterial challenge.

The main function of **macrophages** and **lymphocytes** is to recognise bacteria and then trigger alarm systems which induce a more vigorous host response, eventually leading to huge numbers of PMNs entering the milk. These alarm systems are the inducible defence mechanisms described in the next section.

Inducible Defence Mechanisms

When all else has failed and when bacteria have penetrated the teat canal and overcome the intrinsic defence mechanisms, alarm signals are sent out to the body of the cow requesting 'help'. The response to the alarm is the induced system of mammary defences. It is both highly effective and fascinating in its mechanisms. The various stages will be described in some detail.

The chemotaxin alarm

The macrophages and PMNs (see Table 3.4) already present in the milk recognise and engulf fragments of dead bacteria and their toxins in a process known as **phagocytosis** (Figure 3.2). Phagocytosis in turn leads to the release of various chemical mediators known

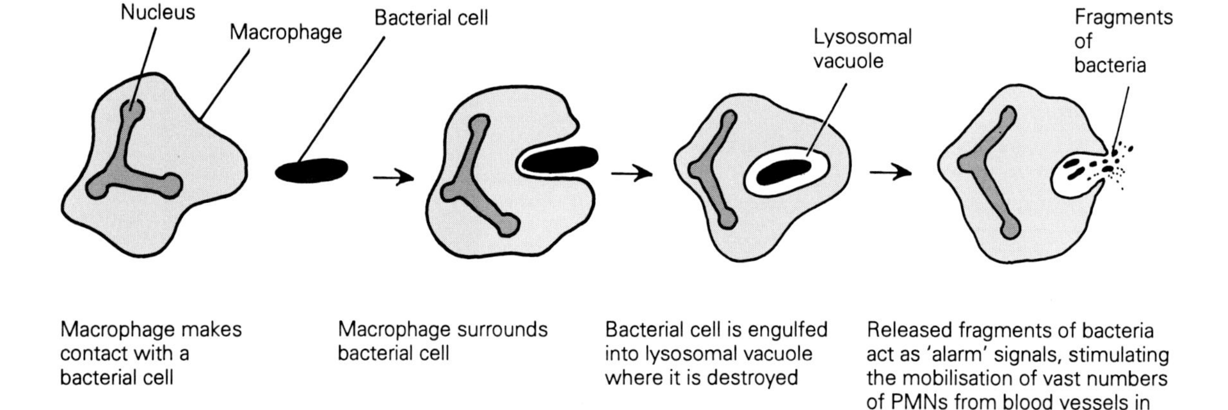

FIGURE 3.2 The process of phagocytosis, in which a macrophage engulfs and destroys a bacterial cell.

collectively as chemotaxins. Specific chemotaxins include chemicals such as interlukin 8 and tumour necrosis factor. It is these chemicals, plus the toxins produced directly from bacteria multiplying within the udder, which act as the alarm system.

Inflammatory response

The principal response to chemotaxins is a massive inflow of PMNs from the capillaries in the teat wall and udder into the cisterns and ducts. This is achieved in a variety of stages (Figure 3.3):

- increased blood flow: blood vessels in the teat wall dilate, thus increasing the blood flow and the supply of PMNs to the affected quarter. This explains why a quarter with an acute mastitis infection becomes palpably hotter.
- **margination**: small carbohydrate projections (selectins) appear on the inner surface of the capillary wall. These attract PMNs towards the sides of the capillaries.
- loosening of endothelial cell junctions: under the influence of specific chemotaxins, the endothelial cells lining the capillaries and the teat and udder cistern literally move apart to facilitate a more rapid passage of PMNs into the infected milk. They close again when the PMNs have passed through.
- **diapedesis**: PMNs squeeze through the walls of the capillaries, across the tissue of the teat wall and udder, through the endothelial lining and into the milk, where they are able to engulf the bacteria.
- damage to epithelial cells: some of the cells lining the teat duct and lactiferous sinuses can be totally destroyed by the toxins produced by *E. coli* infections and this allows further access of PMNs (and serum) into the area of multiplying bacteria. Plate 4.8 shows the inside of a normal teat, which can be compared to the severely inflamed mastitic teat in Plate 4.9. (See page 36.)

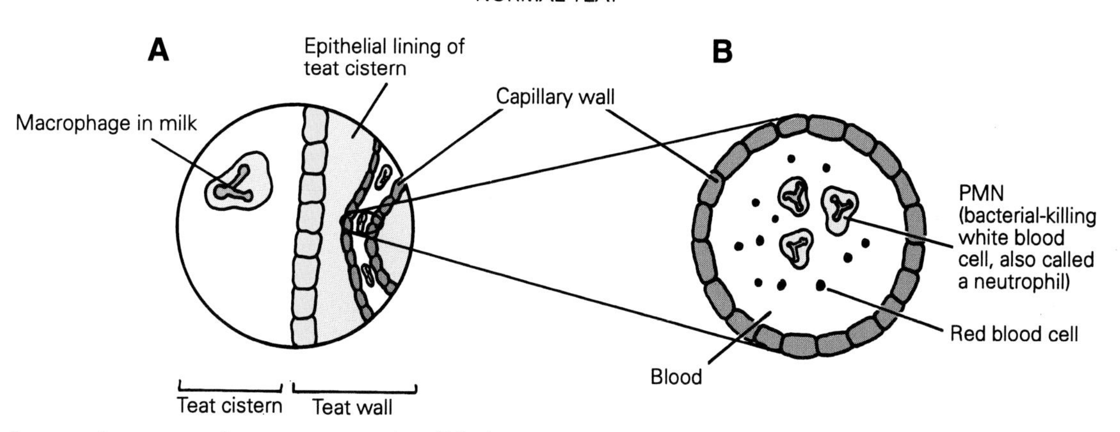

In the normal teat macrophages are present in milk in the teat cistern (A) and PMNs in blood flowing through the capillary (B)

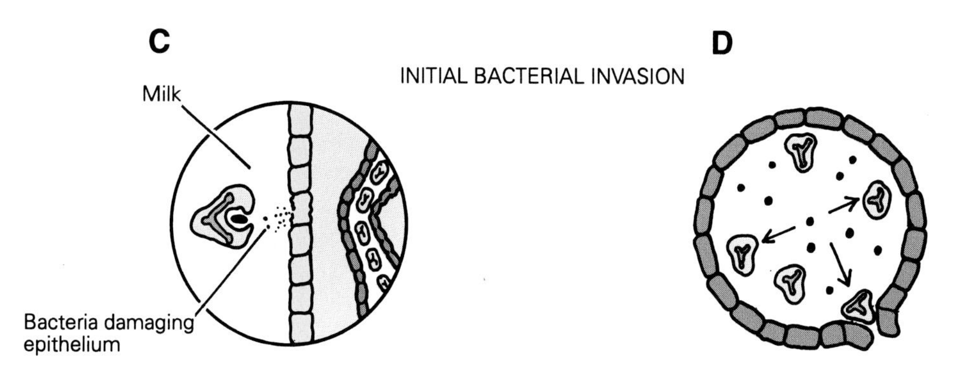

By-products from macrophages and toxins from live and dead bacteria trigger the alarm system (C). The capillary dilates and blood flow increases bringing greater numbers of PMNs. PMNs move towards the capillary wall (margination) and start to squeeze out between the capillary wall cells (diapedesis) (D).

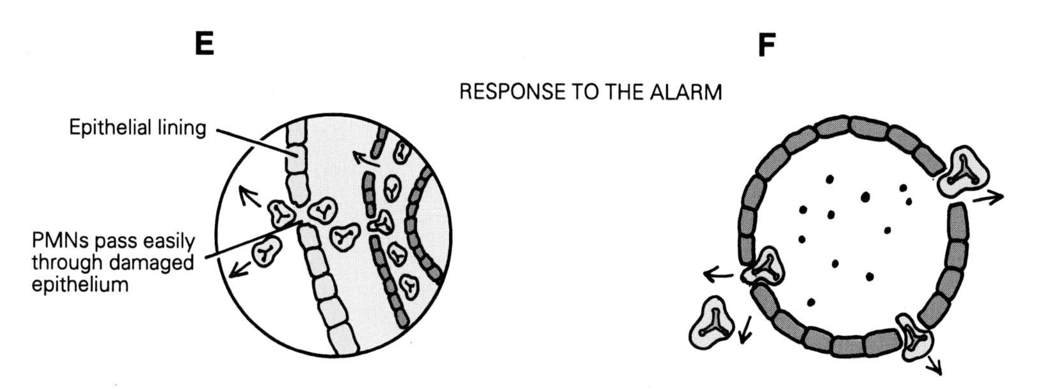

(E) and (F) Huge numbers of PMNs pass into the milk in the teat and udder cistern to produce a massive increase in cell count.They start engulfing and killing bacteria, releasing more by-products which further emphasises the alarm system.

FIGURE 3.3 The response to the alarm signals of bacterial invasion.

- serum ooze from blood vessels: because the junctions between endothelial cells in the capillary walls have opened to allow the passage of PMNs, serum can also flow into the tissues. This produces an uncomfortable swelling of the affected quarter as tissues stretched and dilated by fluid are painful. In acute *E. coli* infections particularly, the leakage of serum is so pronounced that it flows directly into the milk and produces the yellow, watery secretion which is so typical of an acute coliform mastitis. Occasionally serum ooze may even be seen on the skin surface, as in Plate 4.10.

Once they have passed into the milk, the PMNs released in response to the alarm start to engulf whole bacteria (Figure 3.2) and the major part of the bacteria killing process begins. Inside the PMN the bacteria are destroyed by a system involving hydrogen peroxide. The first PMNs to arrive are highly active. They release lysosomal granules from their cytoplasm and this further amplifies the inflammatory response described above. The severity of the inflammation is often such that it persists well after the bacteria have been destroyed. This explains the common finding of a hard, hot and painful quarter with a watery secretion, from which bacteria cannot be cultured. This is almost certainly caused by an acute *E. coli* infection which has been rapidly counteracted by the cow's defence mechanisms.

The increase in the number of cells in milk due to the inflammatory response can be enormous. From a base level of only 100,000 (10^5) per ml, it may increase to as many as 100,000,000 (10^8) per ml in just a few hours. Bacteria are then rapidly eliminated, as shown in Figure 3.4a.

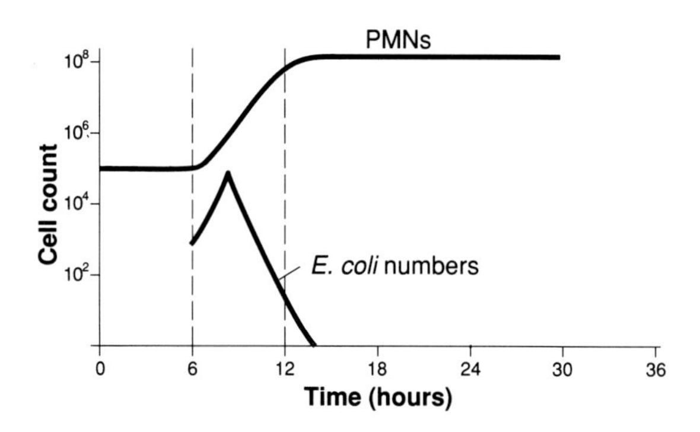

FIGURE 3.4a Good PMN (white cell) response in a mid-lactation cow can lead to rapid elimination of E. coli. *(10)*

The Freshly Calved Cow

The description given above applies to cows which are able to mount a dramatic inflammatory response, producing a hard, hot swollen quarter. Some of these cows may be sick, others less so. There is, of course, an alternative reaction. For some reason, a proportion of freshly calved cows seem unable to mount a PMN response. In such cases, when *E. coli* invade the udder they continue to multiply almost unchecked. In the instance shown in Figure 3.4b as few as 10 organisms (a minute number) may have infected the quarter 12–18 hours

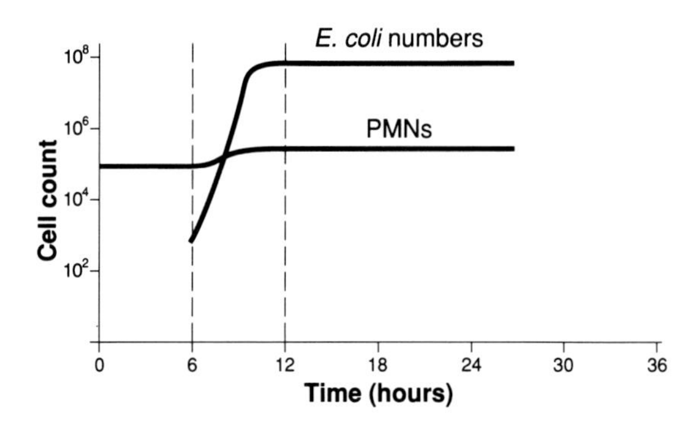

FIGURE 3.4b A poor cellular response seen especially in some freshly calved cows allows E. coli *to multiply to very high numbers in the udder (compare this with the good response shown in FIGURE 3.4a). Provided the cow survives, bacterial numbers may remain high for several days. (10)*

earlier, but because the cow was unable to mount an immune response, the bacteria continued to multiply, with bacterial levels reaching 10^8 (100,000,000) per ml.

Because there is no inflammatory response in such cases, the mastitis may be difficult to detect. The udder may well remain soft and the changes in the milk could be minimal, making it almost indistinguishable from colostrum. However, the cow herself will be very ill, due to the systemic effects of large quantities of **endotoxin** which have been produced by the multiplying *E. coli* bacteria. (Not all bacteria produce endotoxins.) The cow will probably be recumbent and unable to rise, dull and not eating. She may or may not have a temperature (cows with a good inflammatory response invariably have an elevated temperature) and she will probably be shivering, and have profuse, foul-smelling diarrhoea.

Cows which do not die may remain seriously ill for some considerable time. The lipopolysaccharide endotoxin produced by the *E. coli* has a generalised effect on all body organs, leaving the cow in poor condition, dull and with a poor appetite, sometimes for several weeks. There is little that can be done for such cows, since the damage to the udder tissue has already occurred and it is simply time, nursing and tissue regeneration which will effect a recovery.

When phagocytic cells eventually appear in the udder they are often monocytes, a different type of blood cell, which is much less effective than PMNs. It is for this reason that a proportion of cows continue to shed *E. coli* for some considerable time after infection, despite antibiotic therapy. The damage to the lactiferous sinuses and mammary endothelial cells may be so severe that they degenerate, becoming keratinised in appearance, and cease to produce milk for the remainder of the lactation (Figure 3.5). Fortunately the majority of such quarters recover in the following lactation.

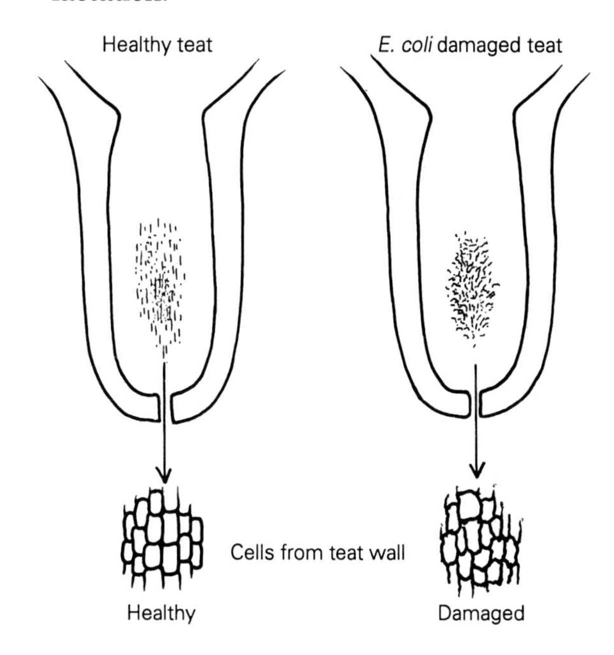

FIGURE 3.5 Damage to teat lining following an acute E. coli *infection.*

The precise reason why some early lactation cows are unable to mount a significant PMN mobilisation and inflammatory response, with such catastrophic effects, remains unclear.

Individual Cow Variation

There is a considerable variation between individual cows in their response to an *E. coli* challenge even in cows at the same stage of lactation. For example, when *E. coli* bacteria were experimentally infused into two different cows:

- 98% were killed within 6 hours in one cow compared to
- only 80% killed within 6 hours in the second cow

Part of this is undoubtedly due to an inherent difference in the rate at which PMNs can kill bacteria. Using test tube experiments it can be shown that PMNs taken from the blood of different cows will kill (or eliminate) *E. coli* at different rates. However, the main difference between cows is the rate at which cells can be mobilised from blood into the teat and udder sinuses.

The Effect of Low Cell Counts

There is a body of opinion which suggests that if somatic cell counts are too low, then cows are more prone to developing the peracute and fatal form of *E. coli* and other types of mastitis. This is unlikely to be the case. The difference between an initial cell count of 50,000 or 250,000 cells per ml is almost insignificant when with clinical mastitis cell counts are likely to rise to 100,000,000 per ml within a few hours. It appears to be the speed at which cells can be mobilised into the udder (rather than the number present initially) which is the critical factor.

Low Selenium and/or Vitamin E.

Macrophages and PMNs engulf bacteria and destroy them. One of the methods of destruction is the release of lysozymes (destructive enzymes) within the PMN vacuole, with the resultant production of hydrogen peroxide. A vacuole is simply a compartment within a cell. The hydrogen peroxide thus produced needs to be destroyed immediately (along with the engulfed bacterium) by the action of glutathione peroxidase, (GSH-PX), a selenium-dependent enzyme. Failure to destroy the hydrogen peroxide can quite rapidly result in the death of the phagocytosing cell itself.

Vitamin E reduces the rate of hydrogen peroxide formation within the PMN and stabilises its cell membranes against its attack, while selenium increases the activity of GSH-PX. Workers in North America have demonstrated a correlation between dietary levels of selenium and vitamin E and mastitis and recommend supplementation of 1000 iu vitamin E per cow per day during the dry period and 400–600 iu per cow per day during lactation. British diets are likely to contain a higher proportion of grass silage and are less likely to be vitamin E deficient. However, one survey carried out in Britain did show that in low mastitis incidence herds there was a correlation between increased cell count and low GSH-PX levels: those herds low in Vitamin E/selenium had higher cell counts. Diets which include large quantities of maize silage should be supplemented with vitamin E.

Reduced PMN Activity in Milk

Unfortunately PMNs are less active in milk than in blood and this is a further reason why peracute mastitis and endotoxic shock may occur under some circumstances. The reduced PMN activity is thought to be associated with a variety of factors, including the following:

- they may become coated with casein which reduces their activity.
- PMNs are unable to distinguish fat and casein globules from bacteria. The globules may be continually engulfed, thereby exhausting PMNs.
- oxygen levels are naturally low in milk. They are reduced even further by bacterial multiplication in mastitic secretions. This limits the ability of the PMN to destroy the phagocytosed bacteria.

When PMNs leave the capillaries they effectively need to take their food stores (glycogen) with them. This is sometimes referred to as 'taking their packed lunch'! Once the food has been exhausted, the PMNs become relatively inactive.

Although the above factors limit the activity of PMNs, the system is still highly effective, probably because of the very large numbers of PMNs present. In fact a cow with acute mastitis may pour so many white cells into the mammary gland that the level of white cells in the blood falls dramatically.

CHAPTER FOUR

Mastitis – Causes, Epidemiology and Control

This chapter examines mastitis in general terms, discusses the organisms involved and reviews control measures.

Mastitis will never be eradicated. There are too many different bacteria involved and many of these are continually present. Antibiotic treatment has varying degrees of effectiveness and, for a variety of reasons, vaccination can only ever produce a partial reduction in incidence. The approach to mastitis must therefore be one of **control** and with increased milk flow rates producing ever higher mastitis susceptibility (see page 18), control will become increasingly important in the future.

Mastitis is commonly referred to under the four following categories:

- Clinical mastitis: an udder infection which can be seen, e.g. by clots in the milk, etc.
- Subclinical mastitis: an udder infection which shows no external changes.
- Acute mastitis: sudden in onset and shows severe signs.
- Chronic mastitis: persists for a long time, but not severe.

DEVELOPMENT OF A NEW INFECTION

To be able to appreciate the importance of the various control measures discussed later, it is first necessary to understand how a new case of mastitis occurs.

Arrival of a Reservoir of Infection

Some of the bacteria which cause mastitis are always present in the environment and are therefore called 'environmental organisms'. For these, 'arrival of a reservoir' simply means a change in environmental conditions, leading to a heavy challenge of infection at some site on the animal. In the case of a mastitis infection the challenge will be at the teat end, which becomes badly soiled with mastitic bacteria.

Other infections (e.g. *Streptococcus agalactiae*) are only present in infected cows and 'arrival of a reservoir' indicates either the purchase of an infected cow or perhaps an infected cow calving down into a herd. In this instance the infection is 'contagious' because it passes from cow to cow.

Transfer of Infection from the Reservoir to the Teat

This will generally occur **between** milkings for environmental organisms, since the first stage in the establishment of a new infection is the transfer of bacteria from the environment to the teat end. However, transfer occurs **during** the milking process for contagious organisms and a vector is needed to carry the bacteria from the infected to the non-infected cow (or infected to non-infected quarter). Examples of vectors include the milker's hands, udder cloths (if the same cloth is used on more than one cow) and the milking machine liner.

Penetration of the Teat Canal

There appear to be two ways in which bacteria commonly penetrate the teat canal:

- growth through the canal. After transfer to the teat end contagious organisms, especially *Staphylococcus aureus* and *Streptococcus agalactiae*, begin their 'attack' on the udder by first establishing a colony at the teat end, i.e. they multiply into large numbers. After colonising the teat end, and probably coincident with this, bacteria literally grow up through the teat canal and move into the teat sinus.
- propulsion through the canal. Pathogens, particularly environmental bacteria such as *E. coli*, are commonly forced through the canal, usually with a reverse flow of milk such as occurs with teat-end impacts (see page 63).

One of the reasons for these differences in the way in which organisms penetrate the teat canal is a variation in their inherent ability to adhere to epithelial surfaces. Contagious organisms such as *Staphylococcus aureus* and *Streptococcus agalactiae* have strong adhesive properties. They are therefore able to stick to surfaces and often become established as chronic infections. The environmental organism *E. coli* has virtually no adhesive properties. Hence transfer into the udder is most commonly associated with reverse flow of milk through the teat canal.

Although penetration of the teat canal has been precisely categorised as by growth for contagious organisms and by propulsion for environmentals, it should be appreciated that the distinction is by no means as precise as this. Clearly a reverse flow of milk will assist movement of contagious pathogens through the canal, while there are occasions (e.g. exposure to high teat-end challenge immediately after milking) when *E. coli* seems to penetrate without reverse flow of milk.

Host Response

Even when bacteria have penetrated the teat canal and entered the udder, establishment of infection is by no means a certainty. There is a variety of ways in which the udder can overcome infection and the effectiveness of these mechanisms can vary enormously between cows (see page 24). There is also a variation in host response to the different organisms, as will be seen later in this chapter when the organisms themselves are discussed.

STRATEGY FOR MASTITIS CONTROL

Having seen how a new case of mastitis is established, it is now possible to define a strategy for control. This can be subdivided into three parts:

1. reduce reservoirs of infection. This means keeping the environment as clean as possible and reducing the number of cows carrying contagious organisms, e.g. by dry cow therapy, post-milking teat disinfection and culling.
2. control spread by vectors. This is particularly important for contagious organisms and is discussed in detail in Chapter six.
3. optimise host defences. The host defence mechanisms were described in Chapter three. Keeping teats and teat ends in good condition are obviously vital components of mastitis control and are once again influenced by milking machine function, which is described in detail in Chapter five.

CONTAGIOUS AND ENVIRONMENTAL ORGANISMS

It is not the purpose of this book to go into precise details of every organism which could cause mastitis: over 200 different organisms have been recorded in scientific literature as being causes of bovine mastitis! However, a basic knowledge of the common infections involved is of value. They can be grouped as follows – organisms in bold type cause the majority of mastitis cases.

Contagious

Staphylococcus aureus
Streptococcus agalactiae
Streptococcus dysgalactiae
Corynebacterium bovis
Mycoplasma

Environmental

Coliforms:
E. coli
Streptococcus uberis

Other Coliforms:
Citrobacter
Enterobacter
Klebsiella
Pseudomonas aeruginosa

Bacillus cereus
Bacillus licheniformis
Pasteurella
Streptococcus faecalis
Fungi
Yeasts

There are a number of other, less common causes of mastitis, which are more difficult to categorise into contagious or environmental. These are listed on page 42. Although it is **possible** for this wide number of organisms to be involved, the majority of mastitis cases are caused by a few common bacteria. Table 4.1 shows the incidence of mastitis infection by different types of organism in 1968 compared with that in 1995. Note the enormous decrease in the percentage of *Staphylococcus aureus* cases and the proportional rise in the percentage of environmental cases (*E. coli* and *Streptococcus uberis*). The overall incidence of mastitis has in fact declined dramatically, from 121 cases per 100 cows per year in 1968, to only 50 cases per 100 cows per year in 1995. This decrease was largely due to the dramatic effects of control measures such as post-milking teat disinfection, dry cow therapy and culling, on contagious mastitis.

Table 4.1 The results of one survey showing the decline in the incidence of contagious mastitis between 1968 and 1995 and the proportional rise in importance of the environmental infections *E. coli* and *Streptococcus uberis* (*11*)

Cases of Clinical Mastitis		
	Percentage	
Type	*1968*	*1995*
Coliforms	5.4	26
Strep. agalactiae	3	—
Staph. aureus	37.5	15.4
Strep. dysgalactiae	20.1	10.8
Strep. uberis	17.7	32
Others	16.3	15.8
No. of cases per cow per year	121	50

Although the **percentage** of environmental cases has increased therefore, from approximately 23% in 1968 to 58% in 1995, this is due to a decline in the contagious infections (*Staphylococcus aureus*, *Streptococcus agalactiae* and *Streptococcus dysgalactiae*), rather than a rise in the number of environmental cases.

As described in earlier sections, these two groups of organisms have marked differences in their epidemiology. In summary, these are:

Epidemiology of Contagious Organisms

- the mammary gland and/or teat skin are reservoirs of infection.

- organisms are transmitted from the carrier cow or quarter to the teats of non-infected cows/quarters during the milking process.
- colonies become established at the teat end and slowly grow through the canal over 1–3 days.
- dry cow therapy (see page 159) and post-milking teat disinfection (see page 97) are important means of control.
- herds with a high incidence of contagious infections often have high cell counts but normal TBCs (see page 138).

Epidemiology of Environmental Organisms

- the environment is the reservoir of infection.
- organisms are transferred from the reservoir to the teats between milkings.
- penetration of the teat canal occurs, for example, by propulsion on the reverse flow of milk.
- dry cow therapy is of no value as environmental infections do not persist subclinically and are not carried from one lactation to the next.
- pre-milking teat disinfection is important in control.
- herds with a high incidence of environmental infections normally have low cell counts but may have high TBCs.

SPECIFIC ORGANISMS CAUSING MASTITIS

This section gives a short description of some of the major organisms causing mastitis, their appearance in culture and the type of mastitis they produce. It is certainly not in any way intended to be a comprehensive guide to the bacteriology of mastitis.

It should also be noted that it is **not consistently possible** to determine the organism producing mastitis from clinical signs alone. While there may be a few classic guidelines – for example, the serum-coloured watery secretion produced by an acute *E. coli* infection – these are by no means consistent. *E. coli* can also cause a very mild mastitis, with a few clots seen at one milking, which will have totally disappeared to give a normal udder at the next milking, or even occasionally a recurrent mastitis with a high cell count.

Staphylococcus aureus

Culture

Haemolytic gram-positive cocci seen as white colonies on blood agar (Plate 4.1). Coagulase positive and sometimes referred to as coagulase positive staphylococci. (Cultures are discussed on page 42.)

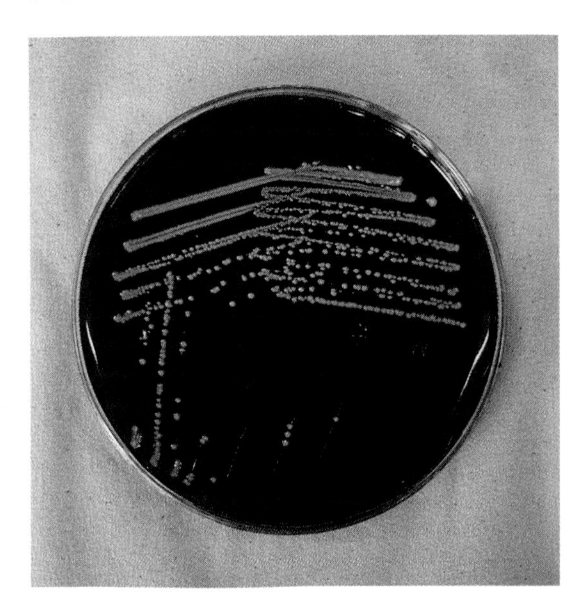

PLATE 4.1 Staphyloccoci growing on a blood agar plate. Each small creamy-white dot represents a colony of staphylococci containing literally millions of bacteria. The slightly lighter ring around the outside of the colony is the ring of haemolysis (broken-down blood).

The primary reservoir for *Staphylococcus aureus* is within the mammary gland. Staphylococci are notoriously difficult to treat and once infection has become established, it is extremely hard to eliminate. Table 4.2 shows that the treatment of clinical cases (that is, where clots are seen) of staphylococcal mastitis with the antibiotic cloxacillin gives only a 25% cure rate. Even treatment of subclinical cases during lactation achieves a bacteriological cure in only 40% of cases.

Table 4.2 Bacteriological cure rates for gram-positive intramammary infections using cloxacillin *(12)*

	Lactation		
Bacteria	*Clinical infection*	*Subclinical infection*	*At drying off*
Staph. aureus	25	40	65
Strep. agalactiae	85	>90	>95
Strep. dysgalactiae	90	>90	>95
Strep. uberis	70	85	85

The reasons for this poor response to treatment are:

- once established within the udder, staphylococci often become 'walled off' by fibrous tissue, which allows only very poor penetration by antibiotics.

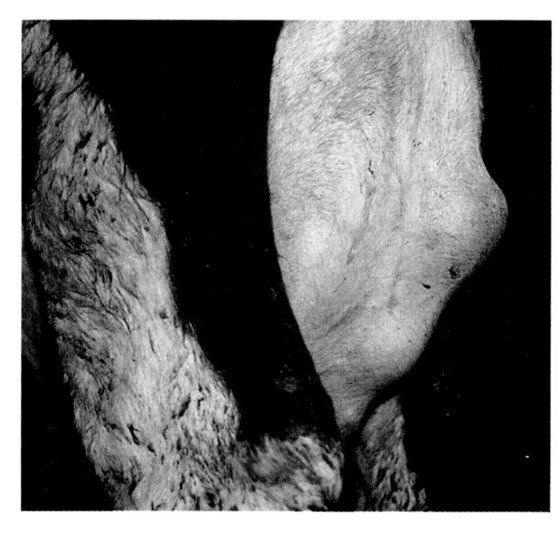

PLATE 4.2 Chronic staphyloccocal mastitis. Note the lumps in the udder.

The cow in Plate 4.2 is obviously affected, with large fibrous lumps protruding from the rear of her udder. She had a cell count of over 3,000,000 cells per ml and suffered from recurrent bouts of mastitis.

- *Staphylococcus aureus* is able to live within cells such as macrophages, PMNs (see page 21) and epithelial cells. Antibiotics can circulate within body fluids but are largely unable to penetrate cells. Those staphylococci which live inside cells are therefore out of the reach of antibiotics.

These two factors also partly explain the very variable cell count and bacterial excretion rates of cows chronically infected with *Staphylococcus aureus*, as shown in Table 4.3.

Table 4.3 Variation in cell count and bacterial excretion rate of a mammary gland infected with *Staphylococcus aureus* (*13*)

Day sampled	*Bacteria/ml*	*Cells (× 1,000/ml)*
1	2,800	880
2	6,000	144
4	7,000	104
5	10,000	896
13	>10,000	152
14	1,200	1,000
15	>10,000	168

These results show very clearly that it would be most unwise to take action (e.g. culling) against a cow on the basis of a single cell count or milk culture result.

The poor response to treatment also emphasises the importance of ensuring that cows do not become infected with *Staphylococcus aureus* and hence the importance of strict hygiene in the milking parlour. Dry cow therapy is vital (see page 159), and although response to treatment is disappointing (see Table 4.2), at least its use lowers the level of infection in a herd and it is one further method of reducing the level of challenge to uninfected cows.

Table 4.4 shows the effects of milking a cow known to be infected with *Staphylococcus aureus* in her udder, on the next cows to be milked.

Table 4.4 Stages in the transfer of *Staphylococcus aureus*, following the milking of an infected cow, to the teats of uninfected cows, using different hygiene routines (*14*)

	% swabs positive for S. aureus from			
Hygiene applied	*Teats before foremilking*	*Teats after foremilking*	*Teats after udder wash*	*Teats post milking*
Water	0	29	63	97
Disinfectant, paper towel and gloves	0	16	39	79

These cows had been tested previously and were shown to be free of *Staphylococcus aureus* on their teat skin. Note how the level of contamination increases each time the teats are handled, and even when strict hygiene is practised (gloves, disinfectant in the wash water and paper towel wipes) contamination still occurs.

Trials have shown that a cow shedding *Staphylococcus aureus* in her milk may infect the teats of the next **6–8 cows to be milked**. The possible level of contamination is therefore enormous and will depend on factors such as the quality of the liners (avoid rough rubber etc.), the initial amount of infection shed and the efficiency of the milking machine. We are, of course, talking about the degree of contamination of the teat skin and not that of actual udder infection. Most infections are killed by post-dipping.

It is clear that if *Staphylococcus aureus* is present in a herd:

- post-milking teat disinfection is **vital**. It will be impossible to prevent the transfer of infection from cow to cow, but much of that infection should be destroyed on the teats before it can penetrate the teat canal.
- ideally, affected cows should be milked last. However, this is often impractical, although milking known infected cows through a separate cluster, which can then be disinfected between uses, will considerably reduce the risk of spread from cow to cow.
- consideration could be given to disinfecting clusters between all cows. This is discussed on page 89.
- teat skin should be maintained in optimum condition. *Staphylococcus aureus* is quite a resistant organism. It can live outside the mammary gland in sites such as udder cloths, the milker's hands and on teat skin. Infections of the teat skin are particularly common if the skin is cracked or chapped, or if it has been damaged by pox virus infections, malfunction of the machine, warts etc. This is another important reason for using a post-milking dip containing an emollient.

Acute Gangrenous Staphylococcal Mastitis

Under certain circumstances which as yet are not fully understood, *Staphylococcus aureus* can cause an acute gangrenous mastitis. This occurs following the production of large amounts of toxin. The condition can be replicated experimentally by removing all immunity from the mammary gland e.g. by infusing anti-bovine leucocyte antiserum, which eliminates all the white cells. In so doing a cow which may have been carrying a chronic *Staphylococcus aureus* infection for months, or even years, may suddenly develop a gangrenous mastitis. Gangrenous mastitis is not caused by a specific acute strain of staphylococcus therefore, but rather by a change in the immune status of the udder.

The classic clinical signs of gangrenous mastitis are shown in Plate 4.3. The skin of the teat and often also of the lower parts of the udder, adjacent to the teat, develops a blue/black discolouration. It will probably be clammy and cold to the touch and may have a slightly sticky feel, due to a surface discharge. In some cases the surface of the skin forms small blisters, as in Plate 4.4. Stripping the teat produces a dark, port-wine coloured secretion (Plate 4.5), often mixed with gas. If the cow is very sick, then the prognosis is hopeless. Even in cows which are not seriously ill there is a risk that at a later date the udder may slough and discharge the affected quarter (as in Plate 4.6). Affected cows are therefore best culled.

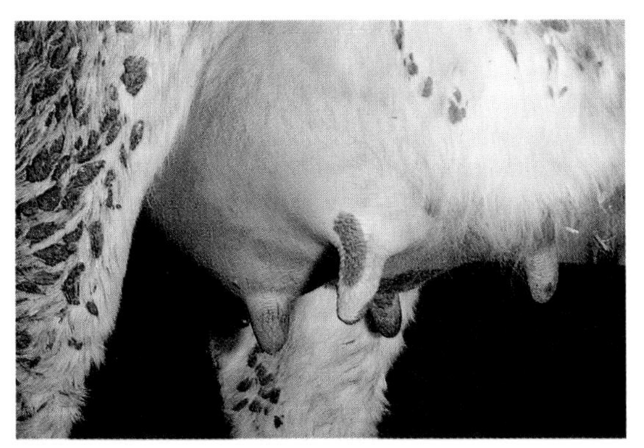

PROFESSOR DAVID WEAVER

PLATE 4.3 Acute gangrenous staphyloccocal mastitis – note the blue/black discolouration. Other organisms such as Bacillus cereus *and occasionally* E. coli *can produce similar changes.*

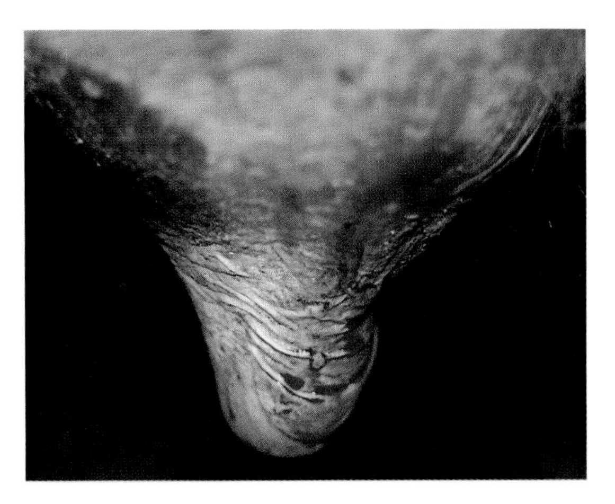

PLATE 4.4 Blistered teat skin associated with gangrenous mastitis.

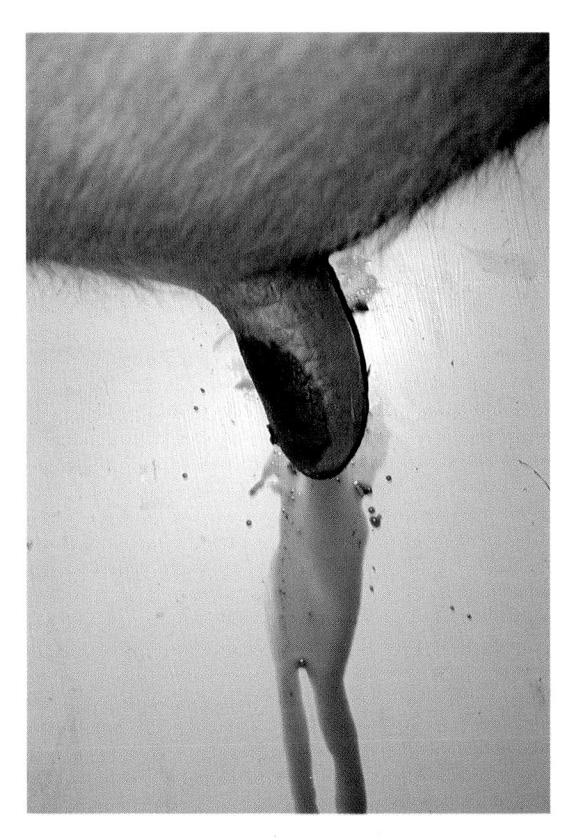

PLATE 4.5 The reddish brown, watery secretion, often mixed with gas, which is characteristic of gangrenous mastitis.

PLATE 4.6 Severe gangrenous mastitis, leading to an udder slough. This cow should be culled.

However, one word of warning. Cows can develop a bruised udder (Plate 4.7), causing blood to accumulate under the skin, resulting in a blue/black discolouration. These cows will be healthy in themselves, their milk will be normal and they will recover without any treatment. They should certainly not be culled as a case of gangrenous mastitis!

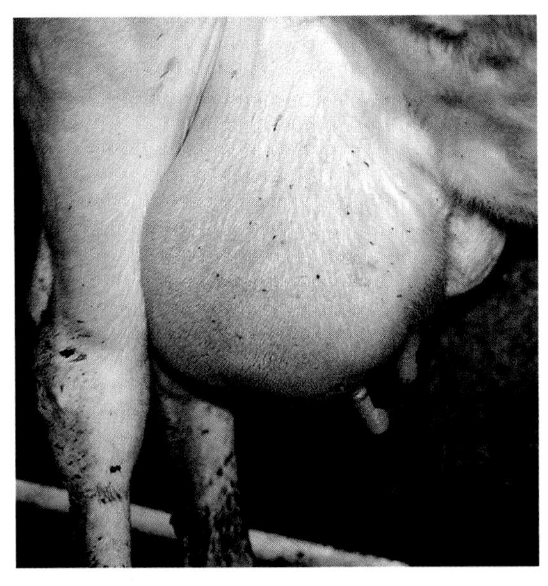

PLATE 4.7 Udder bruising. This must not be confused with gangrenous mastitis.

Coagulase Negative Staphylococci

Coagulase negative staphylococci have been identified as a cause of mastitis. Examples of coagulase negative staphylococci (those which don't form clots in response to the coagulase test and are also non-haemolytic) are: *S. xylosus*, *S. intermedius*, *S. hyicus* and *S. epidermis*. They commonly colonise the teat end and teat canal only. Hence it is difficult to be exactly sure when they are a cause of clinical mastitis. It has been suggested that under some circumstances, they may lead to raised cell counts and subclinical infections. It is very important to discard the first four to six squirts of milk before taking a sample for bacteriology from a mastitic cow. If this is not done, coagulase negative staphylococci may be isolated from the teat canal.

Streptococcus agalactiae

Culture

Gram-positive, non-haemolytic coccus. Very small colonies. Blue appearance on Edwards medium.

Streptococcus agalactiae is a highly contagious cause of mastitis and is easily transmitted from cow to cow during the milking process. Its primary reservoir of infection is in the udder, although it may occasionally colonise the teat canal and even the teat skin, especially if these surfaces are cracked. Its response to antibiotic therapy is good (see Table 4.2) and hence it should be possible to eliminate infection from a herd, provided that careful attention is also paid to the following five control points:

- strict hygiene during milking.
- post-milking teat disinfection.
- dry cow therapy.
- culling of chronic recurrent cases.
- optimum milking machine function.

Milk from infected quarters may contain massive numbers of bacteria, in some cases up to 100,000,000 per ml (10^8). This can lead to elevated and fluctuating TBCs in badly infected herds (see page 134). In such herds a dramatic response can be obtained by a system of **blitz therapy**. This involves treating all cows by infusing antibiotic into each quarter at consecutive milkings. Almost all antibiotics are effective against *Streptococcus agalactiae*, allowing short withdrawal products to be used. Because response to therapy is good and because milk often only needs to be discarded for 24 hours after treatment, this can be an economic procedure in badly infected herds. However, it must be carried out in association with careful attention to the control points listed previously.

The level of *Streptococcus agalactiae* infection in an individual cow is much more closely associated with cell count than with *Staphylococcus aureus* infection. For this reason it is also possible to carry out **partial** blitz therapy where only cows above a certain cell count are treated. This has also produced good results.

Streptococcus agalactiae is thought to penetrate the teat canal by slowly growing through it. Undermilking, which reduces the amount of flushing of the canal, is hence thought to promote establishment of infection. Although mainly an udder pathogen, *Streptococcus agalactiae* can also survive in the environment. For example, it has been shown to persist on milkers' hands, particularly when the hands are badly cracked (as in plate 6.1) and in this way it can be spread from farm to farm.

Streptococcus dysgalactiae

Culture

Gram-positive haemolytic coccus. Very small colonies. Green discolouration of Edwards medium.

Streptococcus dysgalactiae is the third major cause of contagious mastitis. As such it shares many of the properties and control methods applicable to *Staphylococcus aureus* and *Streptococcus dysgalactiae*. However, there are a few specific differences.

Streptococcus dysgalactiae survives well in the environment and has been considered by some to be halfway between contagious and environmental organisms. It is commonly found on teat skin, particularly when the surface integrity is compromised by chaps, cuts, machine damage, pox virus lesions etc. Mammary gland carriers are less important. *Streptococcus dysgalactiae* is also present on the tonsils and hence licking could transmit infection to teats. This could explain why *Streptococcus dysgalactiae* is a common cause of mastitis in heifers and dry cows. Teat irritation associated with flies, or chapping due to cold weather, might encourage an animal to lick its teats and hence transfer infection which gradually colonises the teat canal, until clinical mastitis occurs.

Finally, *Streptococcus dysgalactiae* is commonly found as part of the summer mastitis complex (see Chapter 13) and can be isolated from the carrier fly, the sheep head fly *Hydrotea irritans*.

Mycoplasma species

Culture

Colonies are slow-growing and are said to have a typical 'poached egg' shape when grown on blood agar. *Mycoplasma* needs special culturing facilities and cannot be grown using the techniques described on page 42.

There are two common species of *Mycoplasma* mastitis: *M. bovis* and *M. californicum*. They are highly contagious and can rapidly spread in an infected herd. Response to antibiotics is poor and once identified, infected cows should be milked last, in a separate group and monitored until self-cure has occurred. However, most cows have to be culled. Although infected cows are not clinically ill, infection can lead to a pronounced drop in yield, often referred to as agalactia (meaning 'no milk'). Affected quarters may be swollen and produce only a scant 'gritty' or sandy secretion.

As this is a highly contagious organism, strict attention must be paid to hygiene at milking. This should include pasteurisation of the clusters between cows (see page 86). Both clinically infected and subclinical carriers shed large numbers of the organism.

Escherichia coli

Culture

Haemolytic and non-haemolytic strains of

gram-negative bacilli. Creamy-white mucoid colonies on blood agar.

E. coli is the most prevalent environmental organism causing mastitis. It is present in large numbers in faeces and hence infection occurs primarily in housed cows and particularly when hygiene is poor. Wet and humid conditions are undoubtedly exacerbating factors, since such conditions allow the spread of faecal material and the movement and multiplication of *E. coli*. Important predisposing environmental factors are discussed in more detail in Chapter 8. *E. coli* penetrate the teat canal by propulsion and hence an increased incidence of mastitis is seen with suboptimal machine function or milking techniques which lead to teat-end impacts.

Experimental studies have shown that *E. coli* penetration of the teat canal by no means always causes mastitis. In some cows the only detectable change is a rise in cell count and in bacterial numbers in the gland. In others, very slight damage to the endothelial lining of the teat wall produces just a few white flaky clots which have disappeared by the next milking. The symptoms of typical *E. coli* mastitis are a hard, hot swollen quarter, with a watery discharge. A proportion of cows develop a severe shock reaction and can die within hours. This variation in the response of the cow to invading *E. coli* and the reasons for the wide differences in clinical signs is discussed on page 24.

Unlike *Staphylococcus aureus* and *Streptococcus agalactiae*, *E. coli* does not adhere to the endothelial lining of the teat and udder cisterns. This is probably one reason why chronic carrier cows with recurrent bouts of *E. coli* mastitis are rare. However, occasional cases of chronic *E. coli* do occur and may have to be culled because of persistent, recurrent mastitis and a high cell count.

E. coli toxins

The toxic effects of *E. coli* mastitis are due to the release of an endotoxin, which is a lipopolysaccharide (LPS) derived from the bacterial cell wall. LPS is removed by phagocytosing PMNs (see page 21), which in turn release lysosomal granules, further exacerbating the shock reaction.

Most cases of *E. coli* mastitis are restricted to the teat and gland cistern. Plate 4.8 shows a normal teat in cross-section, with a creamy-pink coloured teat cistern wall. Contrast this with the intense haemorrhage of the endothelial lining of the teat wall in Plate 4.9, which was taken from a cow which died as a result of *E. coli* mastitis.

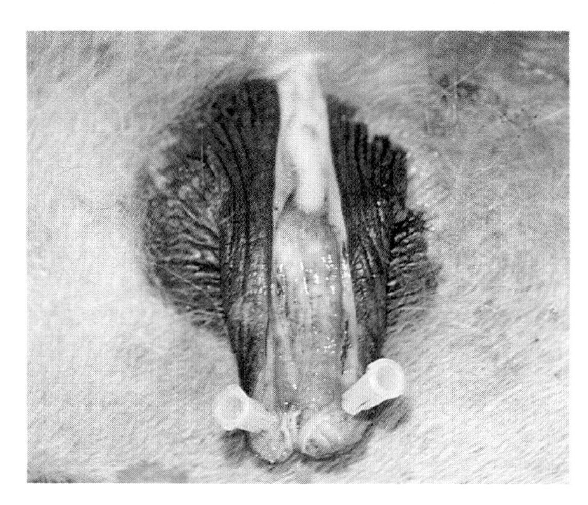

PLATE 4.8 A normal teat in cross-section.

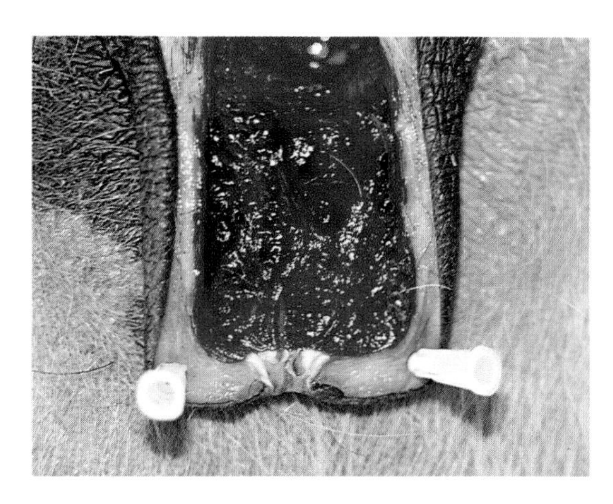

PLATE 4.9 A teat infected with E. coli *mastitis. Note the intense haemorrhage on the teat wall.*

If *E. coli* reaches the smaller ducts and lactiferous sinuses of the main gland, then a massive multiplication of bacteria occurs, and this leads to a severe response in the cow. Sometimes damage to blood vessels is so great that serum ooze is seen on the surface of the udder and teat, as in Plate 4.10. Such cases can

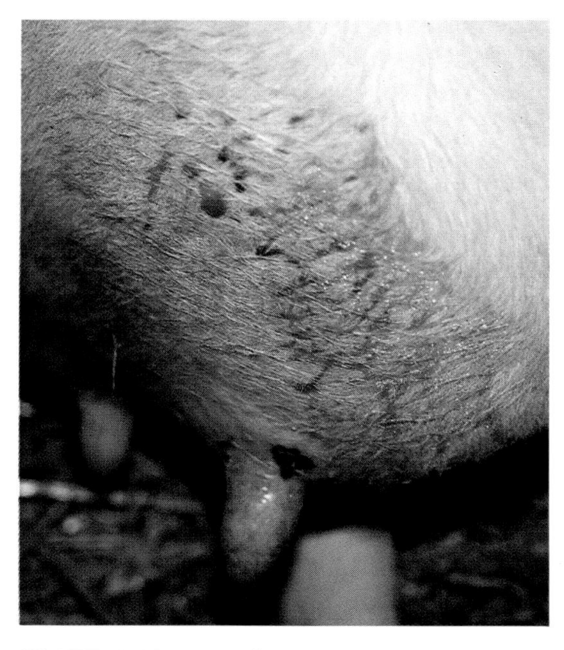

PLATE 4.10 E. coli *mastitis. Damage to blood vessels may be so extensive that serum oozes through the surface of the udder skin, as well as into the mammary gland.*

lead to extensive gangrene and resultant sloughing of udder tissue, as seen in Plate 4.6. This cow should have been culled sooner, although less severely affected cows may discharge smaller areas of gangrenous tissue from the udder and eventually recover. Acute gangrenous mastitis such as this is, of course, quite common in sheep. Many ewes which survive the initial toxaemic attack will slough and discharge the affected half of the udder and recover, with few problems.

E. coli in dry cows

The non-lactating mammary gland is much more resistant to *E. coli*. Although some infections enter, clinical *E. coli* mastitis in dry cows is quite rare. This is thought to be due to the presence of lactoferrin (see page 20). However, those *E. coli* which do gain entry towards the end of the dry period may remain dormant in the udder until lactoferrin becomes inactive. For example, in one trial, *E. coli* was infused into 37 quarters of dry cows 30 days before calving:

23 quarters underwent self-cure and no clinical effects were seen, but
14 quarters developed mastitis after calving.

The importance of this in practical terms is difficult to assess. If we assume that *E. coli* has to be **propelled** through the teat canal (i.e. it is unlikely to **grow** through a closed teat seal), then it seems unlikely that penetration will occur unless:

- the cow is very close to calving and milk is leaking from the udder (i.e. the teat seal has been broken).
- the teat canal is penetrated artificially, e.g. by administering a second course of dry cow therapy or other intramammary antibiotics before calving. If this is carried out, then a preparation effective against *E. coli* should be chosen (see page 150).

Many dry cow antibiotic preparations are not effective against *E. coli* and hence strict attention to hygiene, swabbing the teat end etc., is vital when infusing dry cow tubes.

Variation in strains of E. coli

When a herd outbreak of severe *E. coli* mastitis occurs, it is **unlikely** to be caused by the same strain of *E. coli* in every cow even though clinically this may appear to be the case. There are usually a number of strains involved and the outbreak is caused by some factor relating to the environment and/or milking machine function which is producing a severe challenge. For example, in one survey of 290 isolates from acute *E. coli* mastitis:

82% were typed as 63 different strains of *E. coli*
18% could not be typed

Those strains of *E. coli* which are serum-sensitive (i.e. killed by complement, see page 21) are more likely to be the cause of mastitis. Cow-to-cow transmission, as occurs with contagious mastitis, is unlikely to be important.

Vaccination against E. coli

Vaccination is only possible if an antigen common to all strains of *E. coli* is found.

The toxic effects of *E. coli* are produced by an endotoxin which chemically is a lipopolysaccharide (LPS) derived from the cell wall. Each time one *E. coli* divides into two (and this happens every 20 minutes under the ideal conditions of warm milk within the mammary gland) a certain amount of LPS is released. In addition, when the *E. coli* die, further quantities of LPS are released. Although the LPSs produced by the various strains of *E. coli* are all different, there is one strain known as a 'rough mutant' which produces LPS which has a fragment (known as lipid A) which is common to all LPSs produced by all other strains. This forms the basis of what is known as the J5 vaccine, which has been available in California since 1990 and in the whole of the United States since 1993. Three intramuscular injections of boiled bacteria in an oil adjuvant are given at drying off, 28 days after drying off and within 48 hours of calving. Results of one field trial, in which half of each herd was vaccinated and the other half was left as a control, are shown in Table 4.5.

However, it is interesting that the vaccine does **not** protect against **experimental** challenge by *E. coli* and a good deal of work has attempted to explain this difference. A logical explanation would be that the vaccine in some way alters the method by which *E. coli* infections penetrate the teat canal or become established in the udder. This has still to be clarified.

Table 4.5 Response to J5 *E. coli* vaccine (*15*).

	No. of cows	*Cases of coliform mastitis*	*%*
Vaccinated	233	6	2.6
Non-vaccinated	227	29	12.8

Chronic E. coli infections

In the majority of herds, response to *E. coli* infection is very prompt and the organisms are rapidly eliminated through the natural response of the cows to infection e.g. within 12–36 hours. Cows which mount a poor inflammatory response (see page 24) become very sick and *E. coli* may persist within the udder for 10–14 days, despite the use of antibiotics. This persistent presence of *E. coli* can act as a chronic irritant, leading to hyperplasia (abnormal cell growth) and keratinisation of the gland, and the quarter then dries up. Many such quarters become productive again in the next lactation.

There is also one other form of *E. coli* mastitis, although it is quite rare. In the cows affected clinically, the mastitis may be mild, although cell counts are very high and recurrent bouts of clots occur. Hence these cows need to be culled. The same strain of *E. coli* is continually isolated, despite antibiotic treatment, each time the clots appear. The cause of this is unknown.

Other Coliforms

There is a range of other organisms, in addition to *E. coli*, which fall into the general category of coliform mastitis and may be isolated from time to time. These include:

- *Enterobacter aerogenes*
- *Citrobacter*

- *Klebsiella pneumonia.* This infection has been found in damp, stored sawdust and can produce a severe, toxic mastitis if the sawdust is used for cubicle (free-stall) bedding.
- *Pseudomonas aeruginosa.* Typically *Pseudomonas* originates from contaminated water, found, for example, in udder wash header tanks which are maintained at a low, warm heat and which do not have lids over them or sanitiser added to them. *Pseudomonas* may also be found in water from bore holes. The clinical signs vary enormously, from acute toxic mastitis to chronic recurrent cases. Response to treatment is poor, probably because the organism can live inside cells where it is not accessible to antibiotics. (This is a property which it shares with *Staphylococcus aureus*, see page 31.) Hence, chronically infected cows with high cell counts may have to be culled, since they represent a source of infection to other cows (though it is perhaps not too serious, as only low numbers of organisms are shed).

Streptococcus uberis

Culture

Non-haemolytic, gram-positive coccus. Brown colonies on Edwards plate, due to splitting of aesculin. Some workers categorise all aesculin-splitting streptococci as *Streptococcus uberis*. However, this is incorrect as there are many other examples including *Sreptococcus faecium* and *Sreptococcus bovis*.

Streptococcus uberis is the second most common environmental organism causing mastitis. It is particularly associated with straw yards, where a very high level of infection may occur. Up to 1,000,000 organisms (10^6) per g of straw bedding have been reported. Figure 4.1 shows the correlation between the level of *Streptococcus uberis* per g of straw bedding and incidence of *Streptococcus uberis*

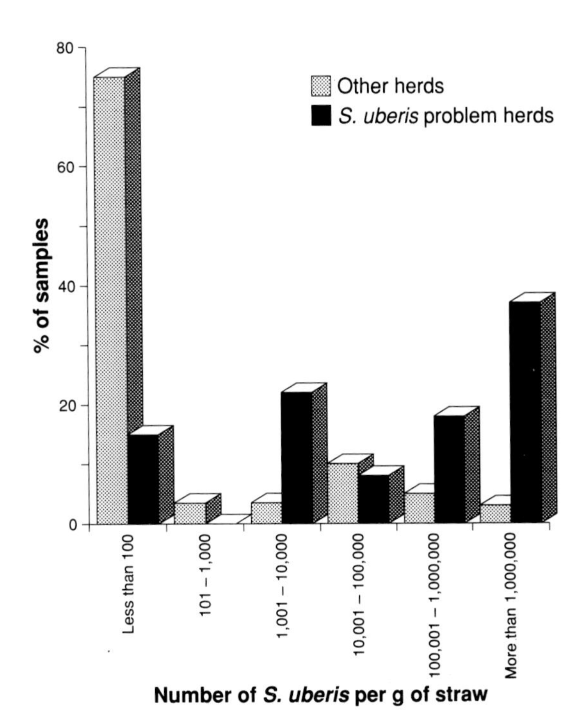

FIGURE 4.1 Levels of Strep. uberis *in straw bedding in* S. uberis *problem herds compared with other herds. (16)*

mastitis in the herds surveyed. Note that the majority of problem herds were associated with high levels of *Streptococcus uberis* in the bedding. The few problem herds with only low levels of infection in the bedding may have been simultaneously affected by teat-end impacts etc. In addition to in the environment, *Streptococcus uberis* can also be found on a wide range of sites on an animal, as shown in Table 4.6.

Table 4.6 Isolation of *Streptococcus uberis* from 53 in-calf heifers kept at pasture under dry sunny conditions (*17*)

Site examined	*No. positive*	*% positive*
Escutcheon	48	90
Legs	46	86
Hind teats	49	92
Lips	12	22
Total	155	73

However, levels in faeces are not particularly high and in this respect *Streptococcus uberis* differs from *E. coli.*

Streptococcus uberis mastitis is often sudden in onset, giving a hard, swollen quarter, clots in the milk and a high, or very high, body temperature. Within the mammary gland opsonisation (coating the bacteria with antibody, see page 21) is poor and hence phagocytosis (engulfing) and destruction of the bacteria by white cells is also poor. Antibiotic therapy is therefore important and in most instances response to treatment is good. A few persistent, recurrent cases do occur however.

Streptococcus uberis is the most common new infection of **dry cows**. The first two weeks and the last two weeks of the dry period are particularly important, though only the last two weeks are significant if dry cow therapy has been administered.

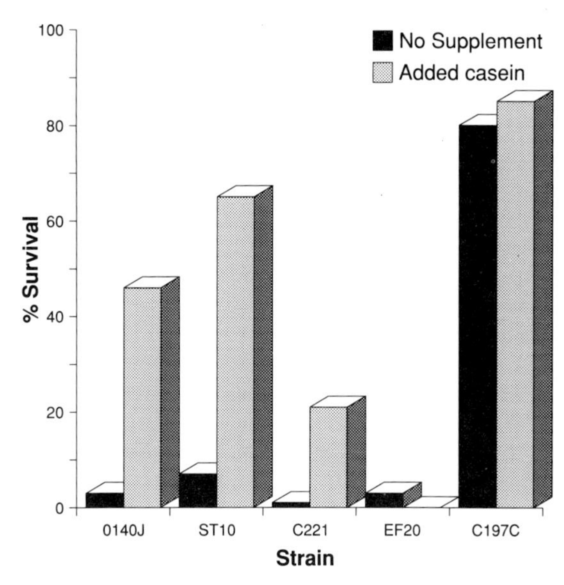

FIGURE 4.2 Phagocytic resistance of Strep. uberis *grown with and without casein. Most strains appear to be protected from phagocytosis in the presence of casein. (18)*

Variation in strains of Streptococcus uberis

Recently it has been shown that what we term *Streptococcus uberis* is not a single organism, but a range of organisms. The various organisms can be differentiated by a technique known as 'DNA fingerprinting' (that is, looking at the genetic material within the cell).

It has been shown (see Figure 4.2) that some strains of *Streptococcus uberis* are much more resistant to phagocytosis by white cells and macrophages than other strains and this is particularly so in the presence of the milk protein, casein. This is especially interesting, because although *Streptococcus uberis* is classically known as an environmental organism, there have been an increasing number of reports which now classify it as one of the more common causes of chronic, recurrent mastitis, which is non-responsive to antibiotics. Could this be the ST10 or C197C strains, which are also highly resistant to phagocytosis?

It has been accepted for quite a time that some cows experimentally infected with *Streptococcus uberis* never recover, despite prolonged treatment with antibiotics, for example, 5–7 days of combined parenteral (by injection) and intramammary therapy. These are also presumably the 'chronic mastitic' strains, which are more closely related to contagious than environmental pathogens and can therefore be transmitted from cow to cow during the milking process.

Other strains of *Streptococcus uberis* appear to confer a degree of immunity: following infection, cows are significantly protected against reinfection with the same strain.

Outbreaks at pasture

The final, somewhat puzzling, aspect of *Streptococcus uberis* mastitis is that outbreaks sometimes occur in cows at pasture. This could be due to contagious transmission of 'chronic' strains, or perhaps to cows transmitting oral infections by licking teats irritated by flies. Alternatively it may be due to the fact that during the summer especially, cows often tend to lie on the same area at night, which could then develop a build-up of infection. This has still to be resolved.

Clinically, *Bacillus species* mastitis is often presented as a hard quarter with white clots. Although sensitivity tests indicate that a wide range of antibiotics should be effective, response to treatment is often disappointing.

Bacillus Species

Culture

Haemolytic gram-positive bacilli with characteristic large, rough,dry and flaky colonies on blood agar.

There are two common *Bacillus* species causing mastitis: *B. cereus* and *B. licheniformis*. Great care must be taken with sampling, since *Bacillus* species can also be a contaminant of the teat canal not associated with mastitis.

B. cereus

This is classically associated with infected brewers' grains and may produce an acute, gangrenous mastitis (see Plates 4.3 to 4.6). It can also occur as a contaminant on the tips of dry cow syringes, especially when they are warmed in a bucket of contaminated water, for ease of administration! Bacteria (including *Pseudomonas*) infused at this stage may persist in the udder to produce acute mastitis at the next calving.

B. licheniformis

This is an environmental organism which can be a cause of mastitis. Cows are particularly vulnerable if they lie on waste silage left beside feed troughs. This is especially the case with maize silage which undergoes more rapid secondary fermentation than other types and therefore produces a warm bed. Poor cubicle comfort, leading to increased numbers of cows lying outside, may also be involved. *B. licheniformis* infections ascending into the vagina may also lead to an increase in endometritis ('whites'), low conception rates and abortions later in pregnancy.

Yeasts, Fungi and Moulds

Culture

Yeasts, fungi and moulds grow slowly on blood agar and are best cultured on Sabouraud's media. Examples include *Candida*, *Prototheca* and *Aspergillus*.

Yeasts and moulds are common environmental organisms. They may cause mastitis if a large number of cows are lying out of the cubicles (free-stalls) or if the milker washes the teats but does not wipe them dry before applying the milking units. This is particularly the case if the water is contaminated and non-sanitised. On farms where wet teats are a problem, heavy contamination of teat skin leads to infection in bulk milk. It may be possible to culture *Candida* species (yeasts), *Aspergillus fumigatus* (moulds) or *Prototheca zopfii* (algae) from both bulk milk and cases of clinical mastitis.

Clinically, the mastitis is most commonly seen as a hard, hot and swollen quarter, with thick white clots. The cow may have an elevated temperature, especially with yeast (*Candida*) mastitis. Treatment with antibiotics is totally ineffective, as yeasts and fungi do not respond to antibiotics. Some success has been reported from infusing 60–100 ml of a mixture of 1.8 g iodine crystals in 2 litres liquid paraffin, plus 23 ml ether, into the quarter, daily for 2–3 days. On each occasion the infusion should be stripped out after 6–8 hours, otherwise the iodine can produce excessive irritation and be a cause of inflammation in itself. Intravenous sodium iodide or oral potassium iodide given concurrently sometimes improves response to treatment.

Uncommon Causes of Mastitis

Scientific literature has implicated over 200 different organisms as causes of bovine mastitis. Only 20 or so have been described so far and of these *Staphylococcus aureus*, *Streptococcus uberis* and *E. coli* are by far the most common. Relatively rare causes of mastitis include:

- *Streptococcus faecalis*: present in faeces and a common contaminant in samples. Could be a cause of mastitis if isolated in pure culture.
- *Leptospira hardjo*: seen in conjunction with abortions and milk drop as part of the leptospirosis complex. Very difficult to culture.
- *Nocardia asteroides:* very hard quarter. Poor response to antibiotics.
- *Streptococcus zooepidemicus*
- *Pasteurella multocida*: environmental organism.
- *Serratia species*: can cause mastitis in both dry and lactating cows. Several species but *S. marcescens* is the most common. Non-pigmented strains are thought to be more pathogenic than pigmented strains.
- *Salmonella*: has possible human health implications.
- *Corynebacterium ulcerans*: has possible human health implications (sore throats).
- *Listeria monocytogenes*: has possible human health implications and has been associated with soft cheese.
- *Mycobacterium smegmatis*
- *Yersinia pseudotuberculosis*: common infection of wild birds, especially starlings.
- *Corynebacterium bovis*: can cause subclinical mastitis and raised cell counts. Has been associated with poor/delayed post-milking teat disinfection. May also be isolated from the teat canal and not associated with mastitis.
- *Haemophilus somnus*.

Specialist texts should be consulted for more details of these organisms. (See further reading section on pages 185–186.)

CULTURING MILK FOR BACTERIA

Because of the great differences in the control methods required for contagious and environmental mastitis, it is obviously important to know which organism(s) you are dealing with before tackling a mastitis problem. This can be achieved by culturing a milk sample on agar plates. These contain special growth media (e.g. blood) for the bacteria.

The teat end is often heavily contaminated by a range of environmental bacteria, but of course they are not necessarily the cause of the mastitis. Some may even penetrate the outer areas of the teat canal. To determine which bacteria are the cause of mastitis, samples must be taken very carefully.

Taking a Milk Sample

The quality of the results (and hence value for money) obtained from submitting a milk sample to the laboratory is to a very large extent determined by the quality of the initial sample. The following procedure should give good results:

1. make sure the operator has clean hands. Wash and dry if necessary, or wear clean gloves.
2. wash and thoroughly dry the teat if it appears dirty.
3. strip out and discard the first 4–6 squirts, thus flushing out non-mastitic bacteria from the teat canal.
4. thoroughly rub the teat end with a swab, until the swab remains clean. Only then should the top be removed from the sample bottle.
5. hold the bottle at an angle of 45° or less and draw out one squirt of milk in a diagonal direction. If the bottle is held vertically there is a much greater risk of dust and debris falling into the sample during stripping. One good

'draw' of milk is sufficient. It is not necessary to fill the bottle.

6. replace the cap and label the bottle with cow identity, quarter, date and farm name.
7. store the sample in the fridge (+4°C) until it can be transported to a laboratory. Ideally, samples should be plated out within 60–90 minutes, as this gives the best results. Storage for up to 72 hours at 4°C is acceptable. Freezing reduces bacterial numbers (especially for coliforms) but can still give useful results.

Laboratory Plating and Incubation

There is a wide range of techniques available, the variation suiting different needs. The method described in the following is currently used by the authors, as it suits their requirements in veterinary practice. Although not the cheapest method, it allows fairly rapid identification of the major groups of mastitic organisms.

1. preheat samples to at least room temperature and preferably 37°C, to break down the fat globules, thus releasing trapped bacteria.
2. use a sterile cotton wool swab to streak the initial plate. A standard 7.0 mm bacteriological loop contains only 0.05 ml milk and such a low volume could miss mastitic organisms which are only present in low numbers in the milk (e.g. less than 20 per ml).
3. plate out onto the following media:
   - sheep blood agar
   - MacKonkey
   - Edwards
   - plus Sabouraud if yeasts and fungi/moulds are suspected (although these organisms will also grow slowly on blood agar).
4. incubate plates at 37°C and examine after 24 hours and again at 48 hours.
5. Gram-stain colonies to determine the organism morphology (structure) and to determine whether it is gram-positive or gram-negative.

   Bacteria are differentiated in many different ways: from their appearance on culture plates, their size, their shape and their reaction to a stain known as a Gram-stain. For example:

   - shape: most bacteria are shaped as either
     - spheres (coccus), e.g. *Staphylococcus* or *Streptococcus*
     - or rods (bacilli), e.g. *Bacillus cereus*
   - Gram-stain: bacteria stain either
     - gram-positive: dark blue, e.g. *Staphylococcus* or *Streptococcus*
     - or gram-negative: red, e.g. *E. coli*, *Pasteurella* or *Pseudomonas*

   Bacilli are often gram-negative (but not always: *Bacillus species* are gram-positive) and cocci are often gram-positive. This distinction between gram-positive and gram-negative bacteria is quite important when comparing the antibiotic sensitivity of different organisms.

   - haemolysis: some bacteria break down blood, to give a ring of haemolysis or 'clearing' of the blood around colonies growing on the agar plate, as seen on Plate 4.1.

6. carry out other useful bacteriological differentiation tests which include:
   - coagulase
   - oxidase
   - catalase

Antibiotic sensitivity testing

A single colony is taken from the original culture plate, added to a bottle of liquid culture medium (broth) and incubated for 6–8 hours, to increase bacterial numbers. It is then poured over the surface of a second agar plate. Bacteria should grow evenly over the surface of this second plate.

Small paper discs, each one containing a different antibiotic, are then placed onto the surface and the plate is incubated for a further 24 hours. If the bacteria grow up to the very edge of the disc (as with the antibiotic on the left of Plate 4.11), then these antibiotics will not be effective in treating the bacteria. The antibiotic has diffused out from the disc onto the agar surrounding the disc

PLATE 4.11 Antibiotic sensitivity plate. Only the antibiotic which causes a zone of inhibition around the disc (right) would be effective in the treatment of this infection.

on the right and prevented bacterial growth. Hence the clear zone of growth inhibition around the outside of the disc. The size of the zone of inhibition does **not** represent the likely efficacy of the antibiotic used in the cow, but rather the concentration of the antibiotic in the disc and the ease with which it can diffuse through the agar. Other factors relating to efficacy of antibiotics are discussed in Chapter 12.

Interpretation of Results

Even when bacteria have been grown and identified, often a mixture of organisms is obtained and there may still be difficulties in determining what is significant. The following are general guidelines:

- a pure culture of any mastitis pathogen – highly probable cause of the mastitis.
- a mixed culture, for example:
  - *Staphylococcus aureus* and *Streptococcus agalactiae*, plus other organisms such as *S. faecalis*, or
  - *E. coli* and *Streptococcus uberis*, plus other organisms

    In the above examples of mixed cultures *Staphylococcus aureus* and *Streptococcus agalactiae*, or *E. coli* and *Streptococcus uberis* are the most probable cause of mastitis and the other organisms are contaminants.
  - a contaminated culture, for example: *E. coli, Bacillus species, Proteus* and *S. faecalis*. These are all environmental organisms. There are so many bacteria present that the sample is obviously heavily contaminated and no useful information can be gained.
- no growth. In any laboratory one might expect as many as 25–35% of samples giving no growth, or no significant growth. This is often frustrating to the herdsman, who may have taken the sample very carefully from a cow obviously affected by clinical mastitis, e.g. with a hard and swollen quarter, or recurrent milk clots.

Possible causes of no growth include:

- *E. coli* infection – host response is so effective that all the bacteria have been destroyed by the time udder changes become visible (see page 24).
- intermittent excretor, e.g. with chronic staphylococcal infection the numbers of bacteria present at different times can fluctuate greatly, i.e. when the sample was taken the organism may only have been present in very low, or undetectable numbers (see Table 4.3).
- residual antibiotic from previous treatment is still present inhibiting bacterial growth in the lab.
- excessive delay between taking the sample and plating it out.
- loop too hot.
- sample volume too small (use a swab).
- unusual organisms, not detectable following standard techniques, e.g. *Mycoplasma* or *Leptospirosis*.
- traumatic or hypersensitivity mastitis, i.e. where no infectious cause is involved. Probably rare.

TOTAL BACTERIAL COUNT, LABORATORY PASTEURISED COUNT AND COLIFORMS

It is unlikely that farmers reading this text will want to carry out their own bacteriology. However, for those involved in laboratory and mastitis investigational work the ability to perform total bacterial, laboratory pasteurised and coliform counts is invaluable.

Total bacterial count (TBC)

This is the total number of living bacteria per ml of milk, sometimes also referred to as TVC, total viable count. Some British dairy companies now require milk with a TBC under 10,000 bacteria per ml, otherwise they impose penalties. The TBC of milk consists of thermodurics, coliforms and many other organisms. Causes of high TBCs in milk and the organisms involved are discussed in detail in Chapter ten.

Laboratory pasteurised count (LPC)

Also known as the thermoduric count (TD), this is a measure of the number of living bacteria present after heating the milk sample. High thermoduric counts are indicative of poor plant washing.

Coliform counts (CC)

The number of coliforms per ml of milk. High coliform counts are usually associated with dirty (faecally contaminated) teats. Values may be increased when there is poor housing and poor pre-milking teat preparation.

Differential count

This involves culturing bulk milk to assess the proportion of different types of mastitis bacteria present. It can help in the investigation of both mastitis and TBC problems. For example, if *Streptococcus agalactiae* is present in bulk milk it could be the cause of both a high TBC and a high somatic cell count. However, a word of warning: the **absence** of *Streptococcus agalactiae* (or any other mastitis organism) from the bulk milk does not mean that it is not present in the herd. It simply means that it has not been cultured on this occasion.

Methodology

It is **vital** that samples remain refrigerated during transport to the laboratory, otherwise TBCs and other counts will increase dramatically. In order to obtain the results carry out the following procedures:

Total bacterial count

Dilute the milk sample 1:1,000 by adding 0.01 ml of milk to 10 ml Ringers solution, then pipette 1.0 ml into a petri dish. Pour on 20 ml milk agar, cooled to 45°C. Allow to solidify then incubate at 37°C for 48 hours. Count the colonies. The TBC is the number of colonies on the plate × 1,000.

Thermoduric count

Heat 10.0 ml milk for 35 minutes at 64°C +/– 0.5°C. Cool and dilute 1:10 with Ringers solution. Pipette 1.0 ml of diluted milk into a petri dish and add 20 ml milk agar cooled to 45°C. Incubate at 37°C for 48 hours and count as above. The thermoduric count is the total number of colonies × 10. Values over 500 per ml suggest a wash-up problem.

Coliform count

Pipette 1.0 ml of undiluted milk into two petri dishes. Pour 20 ml of violet red bile agar (cooled to 45°C) into each. Incubate one plate at 37°C (total coliform count) and the second at 44°C (*E. coli* count). Count the colonies at 24 and 48 hours. Ideally coliform counts should be less than 25 colonies per ml of milk, although values of up to 50 colonies per ml are acceptable.

CHAPTER FIVE

Milking Machines and Mastitis

The milking machine is the dairy farmer's equivalent of a combine harvester. It is a unique piece of equipment as it is the only machine that harvests food from a living animal on a regular basis. Milking machines are used more than any other piece of equipment on the farm. Even so, they are frequently neglected despite the fact that they are responsible for generating the majority of the dairy farmer's income. Milking machines can have an influence both on mastitis and on milk quality particularly TBCs.

HISTORY OF THE MILKING MACHINE

In the early 1800s, a number of pioneers tried to devise a machine capable of milking cows. These people were not farmers, who had little interest in mechanisation while labour was cheap, but plumbers, doctors, inventors and engineers.

The first machine was patented in 1836 and consisted of four metal cannulae connected to a milk pail suspended under the cow. Needless to say this design must have caused considerable damage to the teat ends and also spread of infection from cow to cow. In 1851 two British inventors had the idea of using vacuum to milk cows. In 1895 the Thistle Milking Machine was developed. It used a steam engine to drive a massive vacuum pump and was the first machine to include a device (the pulsator) to relieve constant pressure on the teat.

In the early 1940s only 30% of dairy farmers in England and 10% in the United States were milking by machine. Real development only started in the mid 1940s when the post war shortage of labour stimulated a concerted drive to develop a commercial milking machine that was cost efficient. The development of the milking machine continues even now, with the introduction of more sophisticated electronics to further improve performance.

FUNCTION OF THE MILKING MACHINE

The basic principles of machine milking are identical in all milking systems from the sophisticated rotary parlour down to the milking bale: milk should be removed quickly from the udder with minimal risk to udder health. Milkout occurs by applying reduced pressure, i.e. vacuum to the teat end, which causes the teat canal to open, letting the milk flow out. This is assisted by the oxytocin induced let-down reflex which increases the pressure within the udder (see page 88). A constant vacuum level should be maintained throughout milking. The pulsation system is responsible for ensuring adequate blood circulation around the teat.

In order to understand how the milking machine operates, it is important to know how and where the various components

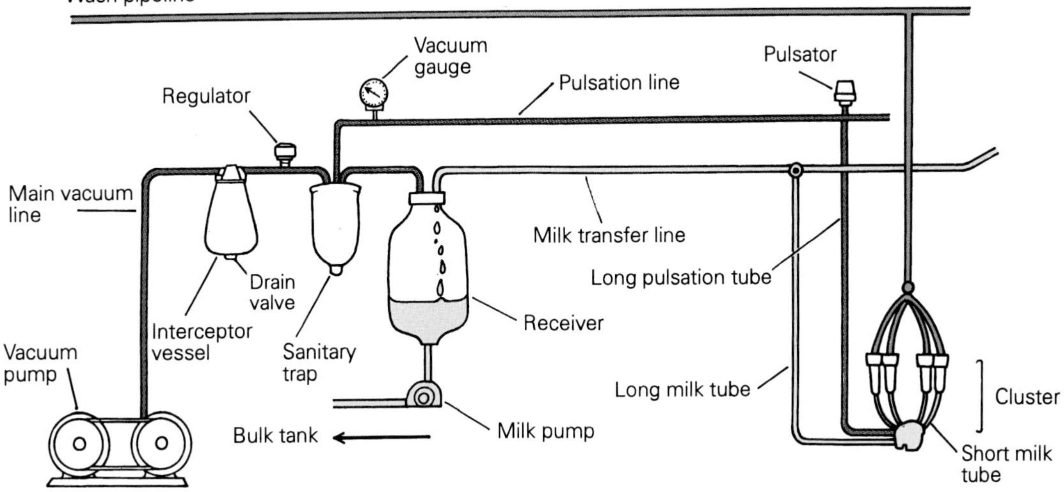

FIGURE 5.1 A simple milking system. There is a huge variety of layouts in milking systems. This depends on factors such as age of the system, whether it has been modified, the amount of space available, manufacturer of the machine etc.

fit into the system. When trying to identify any component, it is advisable to work your way from the vacuum pump forwards so as to avoid confusion. The components of the milking system are described below, and this section should be read in conjunction with Figure 5.1.

Vacuum Pump

This is the heart of the milking plant. It creates vacuum by extracting air from the milking system. Air is removed from all the pipes, jars, claws and liners. Vacuum levels are measured in kilopascals (kPa) or inches of mercury (″Hg). One kPa is equivalent to 0.3 ″Hg or 1.0 ″Hg is equivalent to 3.33 kPa. Vacuum pumps are rated according to the amount of air that they can extract at a set vacuum level, normally 50 kPa (15 ″Hg). This measurement is expressed in litres of displaced air per minute (l/min), and in the United States is expressed in cubic feet per minute (cfm).

The vacuum pump must always be fitted with a belt guard to protect against injury. Plate 5.1 shows a vacuum pump with no belt guard.

PLATE 5.1 The vacuum pump shown has no belt guard to protect against injury and leaves staff at risk from becoming entangled in the belts (A).

The vacuum pump needs to extract more air than is necessary to operate the milking system. This overproduction can be measured and is called the **vacuum reserve**. Vacuum reserve is needed to allow for air admission, such as when units are put on or taken off, and maintains a stable vacuum level during milking. Table 5.1 shows some examples of air admission that occur during milking.

PLATE 5.2 The interceptor vessel with a drain valve.

Table 5.1 Air admission during milking (in litres of air/min) showing the importance of having adequate vacuum reserve if constant vacuum levels are to be maintained throughout milking.

Item	*Air admission (l/min)*
ACRs (per unit)	5–25
Air bleed per unit	4–12
Feeders (per feeder)	5–30
Gates (per gate)	10–42
Liner slip	28–170
Unit attachment	3–225
Unit fall-off (per fall)	570–1400

The amount of vacuum reserve required for a milking plant will depend on the number of milking units and any other equipment that uses vacuum such as ACRs (automatic cluster removers), pneumatic gates, feeders and teat dip sprayers. Allowance should also be made if more than one milker operates the plant. There must be sufficient vacuum reserve present to maintain vacuum stability throughout the whole of milking.

Interceptor Vessel

This is located between the vacuum pump and the sanitary trap. Its function is to prevent any liquids or foreign matter from entering and damaging the pump. There is a drain valve at the base of the interceptor vessel as shown in Plate 5.2.

Balance Tank

This is found in many newer installations and is located between the interceptor vessel and the sanitary trap. It is a large hollow vessel up to 200 litres in capacity that acts as a vacuum reservoir (Plate 5.3). It is designed to improve vacuum stability during milking. Each balance tank has a drain valve at the base. In most installations, the vacuum lines for milking and pulsation feed directly off the balance tank.

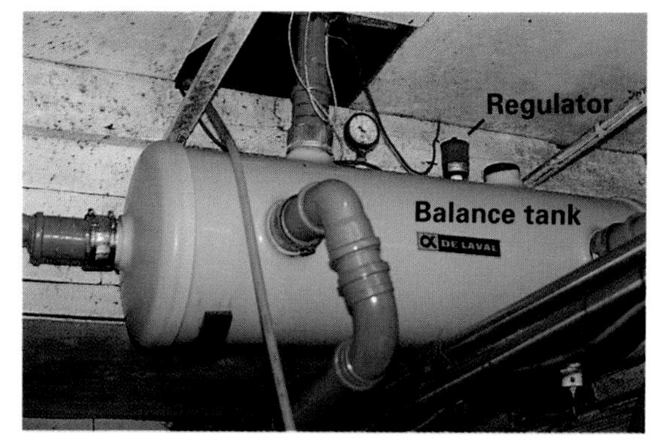

PLATE 5.3 The regulator should be always fitted to the balance tank if one is present in the system.

Regulator

The vacuum pump extracts a **fixed** amount of air from the milking system. However, the demand for vacuum is **variable** depending on how much air enters the system during milking, and so there must be some form of regulation to maintain stability. The vacuum regulator, sometimes called the vacuum controller, is responsible for maintaining vacuum stability.

The regulator leaks air into the system as and when necessary so that a constant pre-set vacuum level is maintained at all times. The regulator should be situated between the vacuum pump and the first cow. It should be sited in a dust-free location where it is easy to clean and inspect when necessary as shown in Plate 5.4.

Many regulators are now fitted with a **clean air system**, as shown in Plate 5.5, that draws air from the external environment. This avoids dust contaminating the regulator and entering the milking system.

PLATE 5.4 The regulator should always be sited in a clean area free from dust.

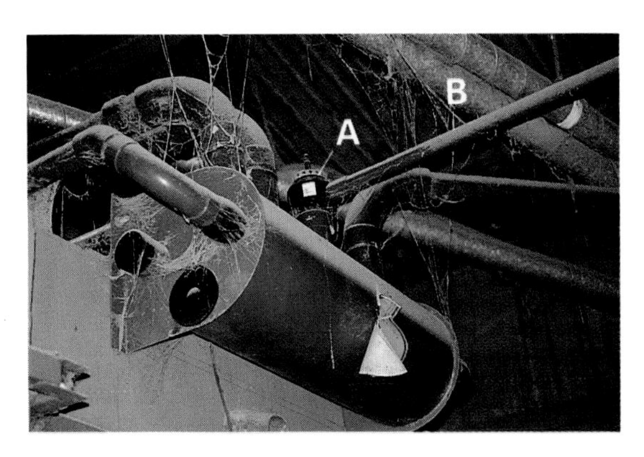

PLATE 5.5 Regulators fitted with a clean air system are surrounded by a case (A). A pipe (B) feeds air from the external environment through the regulator and into the milking system. Although the surrounding area is dirty the system should prevent this dirt from entering the pipelines. A clean air system is an optional extra that may be available for some forms of servo regulators.

The regulator filter should be cleaned regularly because if it becomes blocked it may be unable to respond rapidly to vacuum changes in the system. This will result in poor vacuum stability and increase the risk of fluctuations and thereby new infections. The regulator should be located on the balance tank, if one is present in the system (see Plate 5.3). The balance tank acts as a reservoir for vacuum so there should be more stability and less turbulence at this point than on the pipes that feed in and out of the tank.

Regulators are rated according to the amount of air that they can admit into the system. Regulator capacity is measured in litres of air per minute (cfm – cubic feet/minute – in the United States) and should equal the capacity of the vacuum pump. Any defect in regulator function may result in vacuum fluctuations during milking. The regulator should always be leaking air into the system. If this does not occur it indicates that the plant is unable to maintain a stable vacuum level.

There are three types of regulator: **weight**, **spring** and **servo-operated**. Weight and spring-operated regulators measure the vacuum level at the place where the air is admitted. For this reason, they do not respond very quickly to pressure changes. Servo or **power-operated** regulators have a vacuum sensor fitted away from the air inlet valve and can correct vacuum fluctuations within milliseconds. Servo regulators are highly efficient. A spare regulator should always be kept on the farm in the event of breakdown.

Sanitary Trap

This is located at the junction of the milk and air systems and is shown in Plate 5.6. Its function is to prevent any milk or liquids from entering those pipelines that carry only air, such as the line to the vacuum pump or to the pulsation system.

The sanitary trap should be made of a transparent material such as glass or pyrex and be located where the milker can keep an eye on it during milking. The sanitary trap is fitted with a floating ball valve so that if fluids build up, the valve rises and shuts off the vacuum supply. This causes all the milking units to fall off the cows, and in so doing protects the vacuum pump and airliner from becoming contaminated with milk.

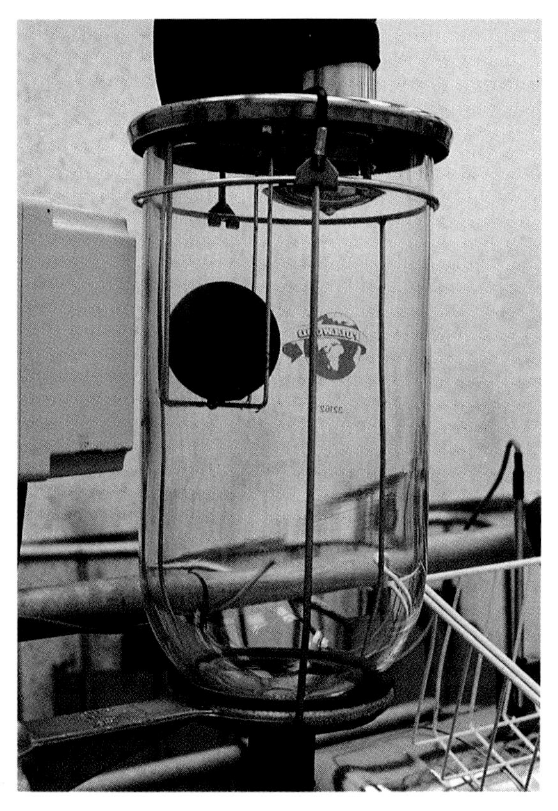

PLATE 5.6 Sanitary trap with a ball shut-off valve to stop liquids entering the airlines.

Vacuum Gauge

This should be located in the parlour so that it is visible to the milker throughout milking. It may be located close to or on top of the sanitary trap. Gauges should be large enough to be read anywhere in the parlour. (See plate 5.7.)

PLATE 5.7 The vacuum gauge must be easy to read, as above, and visible throughout milking.

Pipelines

These carry either air, milk or a mixture of the two. There should be as few bends as possible and no constrictions, as these will interfere with air or liquid movement. The free passage of air and liquids helps maintain vacuum stability. Dead ends must be avoided as they are difficult to clean and can lead to TBC problems (see page 135). Pipelines that carry milk must be made of material, such as stainless steel or glass, that can be cleaned and disinfected.

Large bore pipes are now fitted to most new milking installations. The effect of pipe size on internal volume is shown in Figure 5.2. Four pipe sizes filled with an equal volume of milk are compared. The extra space in the larger bore pipe allows better movement of milk and air compared to that in the narrow pipe which is flooded with milk. Large bore

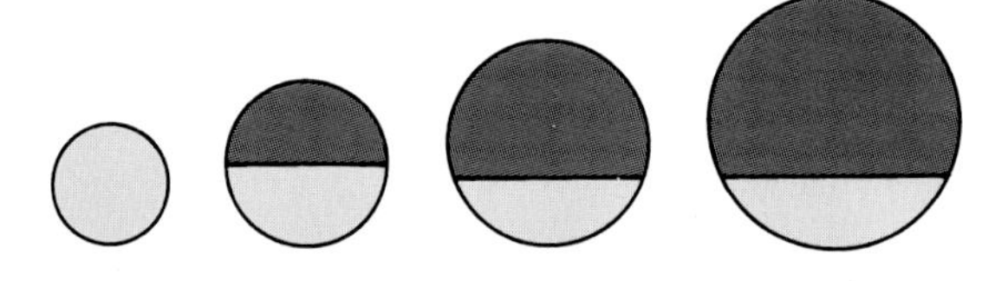

FIGURE 5.2 The same volume of milk shown in pipes of different sizes.

pipes are more difficult to clean however, and need more hot water plus air injectors to produce a swirling effect that will clean the entire internal surface of the pipe (see page 66).

Cluster

This consists of a clawpiece and four teat cups each with its shell and liner plus a short milk and a short pulsation tube. These are connected to the long milk and long pulsation tubes through the clawpiece (Figure 5.3). Milk is removed

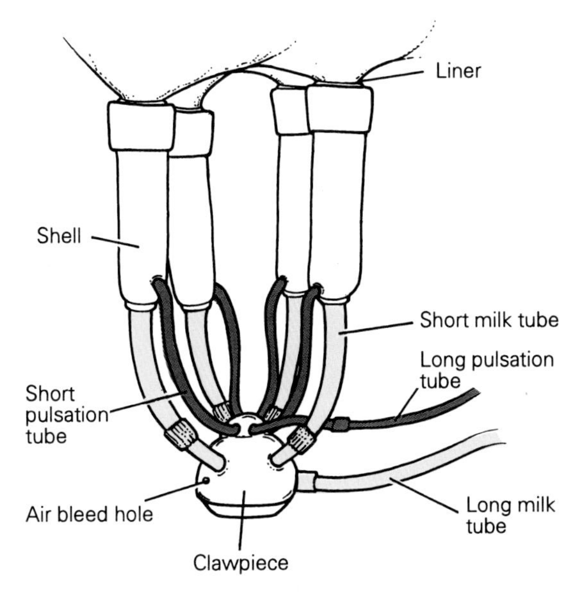

FIGURE 5.3 The cluster consists of a clawpiece and four teat cups each with its own shell and liner, short milk and short pulsation tube.

from the udder into the liner, then through the short milk tube, into the clawpiece and out through the long milk tube. An **air bleed hole** is fitted to each clawpiece. This leaks atmospheric air during milking to assist milk flow away from the udder. Air bleed holes admit between 4 and 12 litres of air per minute. The long milk tube is connected to either a recorder jar or the milk transfer line.

It is important that milk is removed swiftly from the udder. This will stop flooding in the clawpiece and the short milk tubes. Flooding means that milk from one quarter could pass to any of the other three teats thus allowing **cross-contamination** between quarters. Flooding also leads to vacuum fluctuation. For this reason the short milk tube, the clawpiece and the long milk tubes should be of sufficient size to make rapid milk removal possible.

Many years ago clawpiece capacity was as low as 50 cc. However, as milk flowrates have increased (see page 2) a capacity of up to 500 cc is commonplace today. The diameter of the short air and milk tubes has also increased. The difference in volume between a 13 mm and 16 mm long milk tube is 50%, and this can have a marked effect on vacuum stability at the teat end.

Milk flows through the short milk tube, clawpiece, long milk tube, into the recorder jar (if fitted), down the milk transfer line (this may be called the long milk line) and into the receiver vessel. The milk transfer line should be gently sloped to assist the passage of milk to the receiver vessel.

Receiver Vessel

This vessel (see Plate 5.8) receives milk from one or more milk transfer lines. It may be made from either glass or

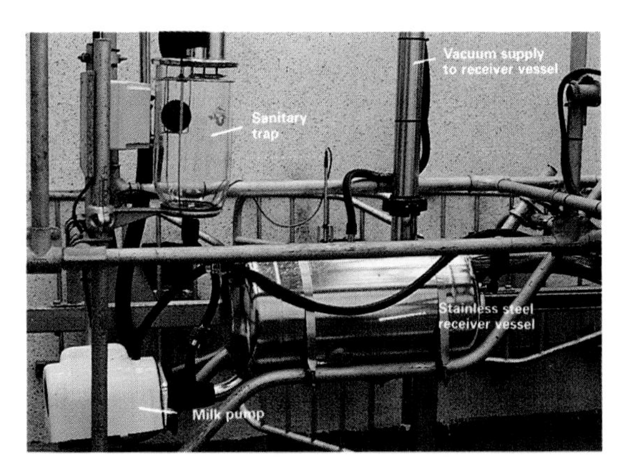

PLATE 5.8 A stainless steel receiver vessel is resistant to breakage but difficult to inspect.

stainless steel. It can be located in the milking parlour or in the dairy. When milk builds up in the receiver it triggers sensors to start the **milk pump** which is connected to the base of the receiver. The milk pump then pushes milk from the receiver vessel under atmospheric pressure into the bulk tank, as shown in Plate 5.9. The receiver is therefore the 'break' between the vacuum system and the outside atmosphere.

Milking systems can be divided into two types depending on the level of the milk transfer line in relation to the cow: if the milk transfer line is below the level of the udder, then the system is called a **low line**, and if above the udder a **high line** (see Plates 5.10a and b). High line systems need to operate at a higher vacuum level as they have to physically 'lift' milk from the udder into the milk transfer line. Milk should not be lifted more than 2 m above the level of a standing cow.

PLATE 5.10a Low line system where the milk transfer line is below the level of the udder.

PLATE 5.9 The milk pump pushes milk from the receiver vessel under atmospheric pressure into the bulk tank.

PLATE 5.10b High line system where the milk has to be 'lifted' above the udder.

Recorder Jar

This is a vessel that holds and stores milk from an individual cow in the parlour. It allows the milker to see how much milk each cow has given. Milk travels from the recorder jar to the receiver vessel along the milk transfer line. In some systems there are no recorder jars and milk is released from the long milk tube directly into the milk transfer line, and then into the receiver vessel. This type of milking is called a **direct to line** system.

Automatic Cluster Removers

More commonly known as ACRs, these are shown in Plate 5.11. They remove the cluster automatically once the cow is milked out. Milk flow is measured by a sensor fitted in the long milk tube. When milk flow falls below a certain pre-set level the vacuum supply to the cluster is shut off. Air enters into the claw through the air bleed hole making the vacuum level drop, and a cord removes the unit from the udder.

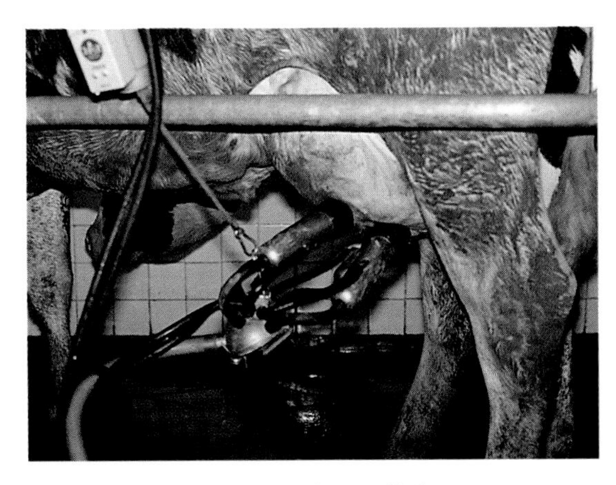

PLATE 5.11 ACRs shut off the vacuum to the cluster at the end of milking. To prevent the unit falling on the floor and becoming contaminated, a cord is activated, which gently lifts the cluster off the cow.

PULSATION

This is responsible for maintaining blood circulation around the teat and causes the liner to open and collapse. The liner should be able to open fully and collapse completely below the teat with full and free movement. This is achieved by alternating atmospheric air and vacuum in the pulsation chamber as shown in Figure 5.4.

Pulsation Chamber

This is the space between the liner and the teatcup shell. Air is extracted from the pulsation chamber through the short and long pulsation tubes. These join on the top of the clawpiece as shown in Figure 5.3. Note there is **no** connection between the milk and pulsation lines on the clawpiece.

During milking, vacuum is constantly applied to the base of the teat. When atmospheric air enters the pulsation chamber it forces the liner to collapse around the teat. This happens because the pressure on the outside of the liner is greater than that inside. When this occurs, milk flow stops and blood circulates around the teat. This is called the **'massage'** phase.

When the atmospheric air is 'sucked' out of the pulsation chamber and replaced with vacuum, the liner is 'pulled' open. This occurs as there is no pressure difference between the pulsation chamber and the inside of the liner and so the liner opens under its own elasticity. The elasticity of the liner is therefore very important. When the liner opens, milk flows away from the udder and this is called the **'milkout'** phase.

One complete liner movement is called a **pulsation cycle**. Each pulsation cycle can be divided into four phases – a, b, c and d, as shown below:

Table 5.2 The pulsation cycle. (*19*)

Phase	*Liner action on the teat*
a	opening
b or milkout	open with full milk flow
c	closing
d or massage	closed: milk flow stops, blood circulates

Each pulsation cycle can be traced onto a graph to show the pressure changes that occur inside the pulsation chamber as shown in Figure 5.4. It must be remembered that this is not pulsation. It is just a graphical representation of the pressure changes that are occurring within the pulsation chamber during a pulsation cycle. Pulsation refers to the actual liner movement around the teat.

Pulsation Rate and Ratio

The **pulsation rate** is the number of pulsation cycles per minute, and the rate is normally between 50 and 60 cycles per minute. The **pulsation ratio** refers to the length of milkout (a + b) compared to massage (c + d) and is expressed as a percentage. A pulsation ratio of 60/40 refers to 60% of the cycle as milkout and 40% as massage.

$$\text{Pulsation ratio} = \frac{a + b}{c + d} \times 100\%$$

To make matters even more confusing there are two different forms of pulsation: single and dual.

Single

Single or simultaneous pulsation occurs when the movement of all four liners in a cluster acts in unison. Single pulsation may be referred to as 4 × 0 pulsation. All four quarters are milked out at the same time and then massaged at the same time. This results in 'slugs' of milk flowing away from the udder.

Dual

Dual or alternate pulsation is where the movement of two liners alternate with the other two. Dual pulsation may be referred to as 2 × 2 pulsation. So while two quarters are being milked, the opposite two are being massaged. This results in a continuous flow of milk away from the udder. Dual pulsation tends to be more efficient than single pulsation as there are no surges of milk or air flow in the system during milkout. In plants with single pulsation there is usually only one long pulsation tube, whereas with dual pulsation there must always be two.

PULSATORS

A pulsator is a device that alternates vacuum and atmospheric air in the pulsation chamber.

There are two different types of pulsators: individual and master.

Master Pulsation

This is controlled electronically and regulates the pulsation throughout the parlour. Each cluster often has its own 'slave' controlled by the master. They are designed to give uniform pulsation no matter where a cow is milked in the parlour. If problems occur with master pulsation then all milking units will be equally affected.

Individual Pulsation

Individual pulsation refers to a system where each milking unit has its own pulsator. All the pulsators operate independently of each other and so cows may receive different pulsation rates and ratios depending on where they are milked in the parlour. If a problem occurs with an individual pulsator, then only that milking unit will be affected. Individual pulsators are expensive and many manufacturers are now ceasing production because of the benefits of master pulsation.

Liner

This is the only piece of the milking machine that comes into direct contact with the cow. Liners are made from complex rubber or silicone material and have a mouthpiece, a barrel and may have an integrated or separate short milk tube as shown in Figure 5.5. Liners have a limited useful life.

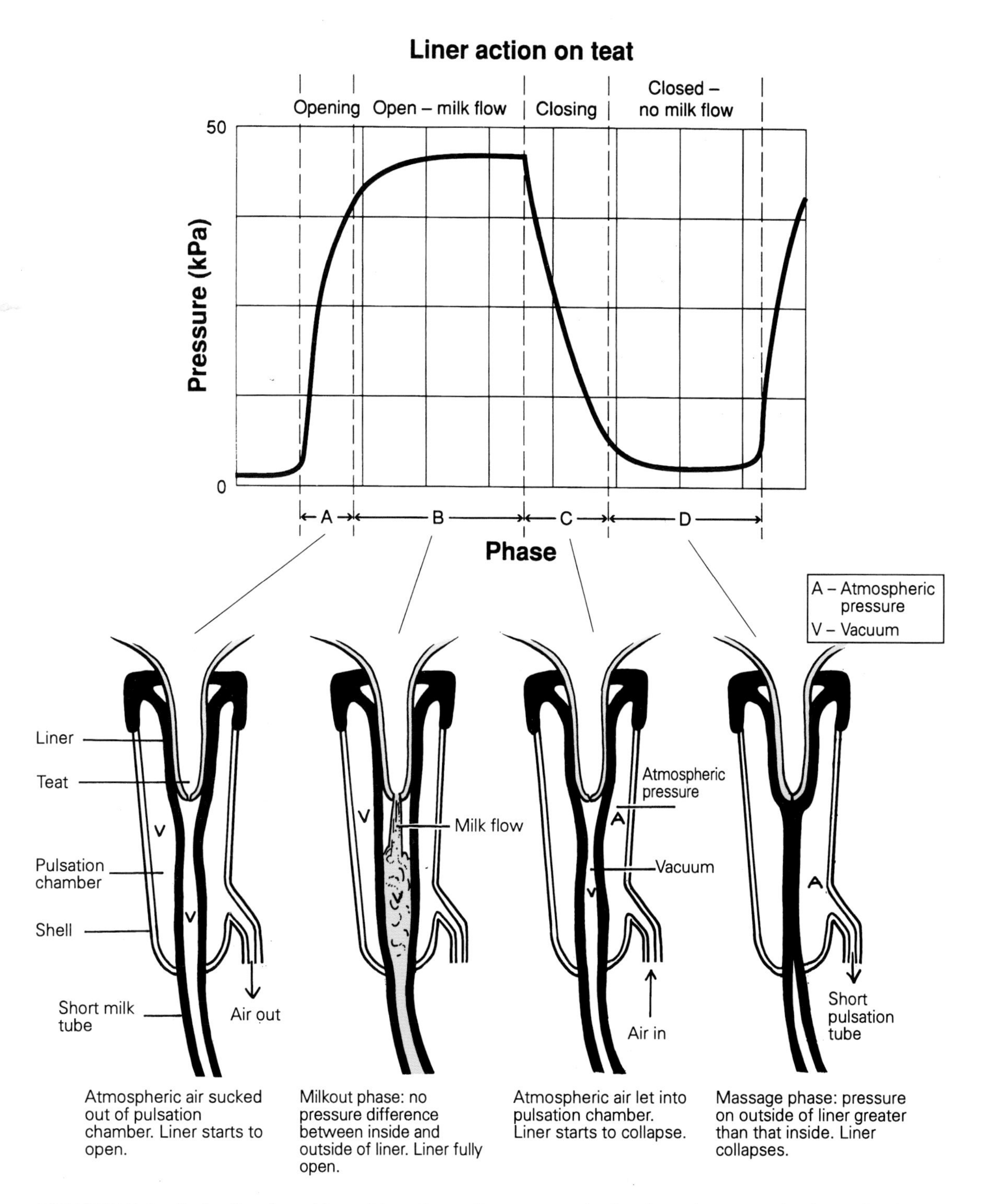

FIGURE 5.4 Pulsation is achieved by alternating atmospheric air and vacuum in the pulsation chamber. Each pulsation cycle can be traced onto a graph (above) to show the pressure changes that occur inside the pulsation chamber.

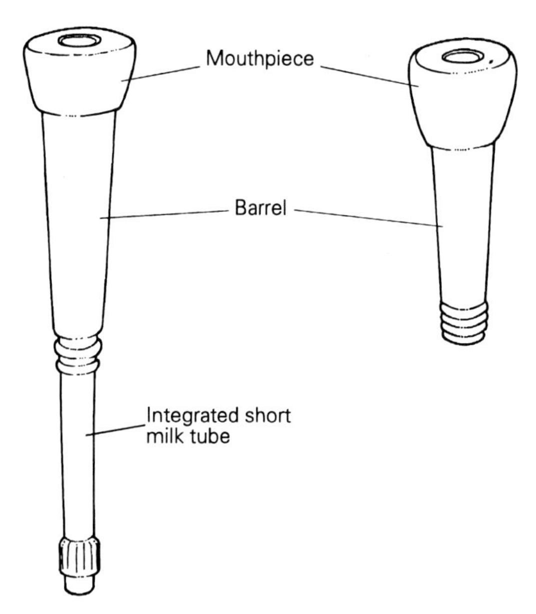

FIGURE 5.5 Liners are made from complex rubber or silicone material and have a mouthpiece, a barrel and may have an integrated or separate short milk tube.

The majority of rubber liners are expected to last for 2,500 milkings or six months, whichever comes first. Silicone liners have a much longer life of up to 10,000 milkings but are more expensive.

Liners should meet the following requirements:

- have a soft, flexible mouthpiece that forms an airtight seal with the base of the teat, adjacent to the udder. This will minimise liner slip and unit fall-off.
- have a barrel long enough to allow the liner to collapse fully around the base of the teat.
- be easy to clean.
- provide a rapid milkout with minimal teat injury.

Liners eventually lose their elasticity and become collapsed as shown in Figure 5.6. This occurs as liners always open and close in the same plane and explains why 'wedging' is sometimes seen at the teat end on some cows after unit removal (see page 180). When liners become collapsed it takes longer for them to open and so milking time will be increased. Many milkers have noticed that milking time is reduced as soon as a set of collapsed liners is replaced. Chemicals, especially chlorine products, will denature rubber and so reduce liner life. A rough inner surface of the liner may abrade teat skin and will certainly harbour bacteria. This will increase the risk of mastitis transmission and may affect TBCs.

Liner usage can easily be checked by using the formula below. The number of cows refers to the total number of cows in the herd, i.e. milking and dry.

Liner usage (no. of milkings) =

$$\frac{\text{No. of cows} \times 2^{*} \times \text{liner life in days}}{\text{No. of milking units}}$$

* Change to 3 if milking three times a day.

For example, in a herd of 160 cows milked twice a day through an 8 x 16 parlour (8 x 16 refers to 8 milking units with 16 cow standings) and where liners are replaced twice a year, liners will have milked 7,320 cows before being replaced.

Liner usage =

$$\frac{160 \text{ cows} \times 2 \times 183 \text{ days}}{8 \text{ milking units}} =$$

7,320 milkings per liner

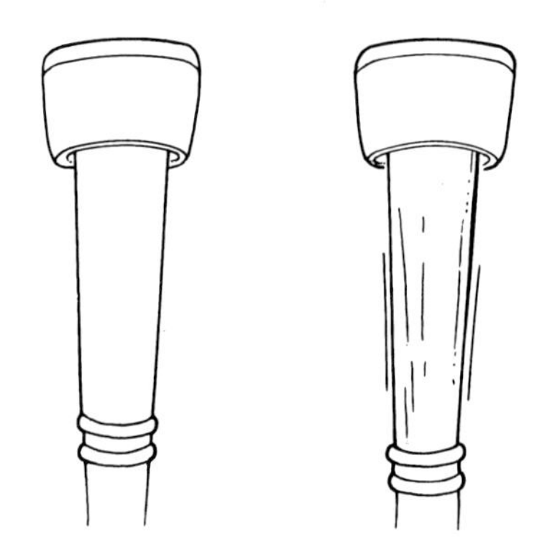

FIGURE 5.6 Liners deteriorate with age, becoming collapsed and losing their elasticity as shown in the liner on the right.

In this case the rubber liners have exceeded their useful life of 2,500 milkings (check the manufacturer's recommendations for useful life of the type of liner you are using) and need to be changed more frequently. The frequency of change can be worked out using the following formula:

Liner life in days (ie. frequency of change required) =

$$\frac{2{,}500^{*} \times \text{no. of milking units}}{\text{No. of cows} \times 2^{\dagger}}$$

* Change if manufacturer's recommendations are different.
† Change to 3 if milking three times a day.

So for this herd the liner life in days =

$$\frac{2{,}500 \times 8}{160 \times 2} = 62.5 \text{ days or 2 months}$$

The frequency of liner change can be checked quickly using the liner life charts in the Appendix. All the information required for this is the herd size and the number of milking units.

Liner choice is very important. The correct liner should be chosen to fit the teatcup shell. It must be able to collapse fully around the base of the teat otherwise blood will be unable to circulate and this may lead to teat-end oedema (see page 180) and other teat-end damage. Once liners become worn and rough, not only can they cause damage to the teat end but they also become more difficult to clean (see Plate 5.12).

PLATE 5.12 Worn, rough liners are difficult to clean and may harbour mastitis bacteria.

Liner Shields

These may be fitted to the base of the liner barrel as shown in Figure 5.7. Their function is to reduce the effect of any impact forces (see page 63).

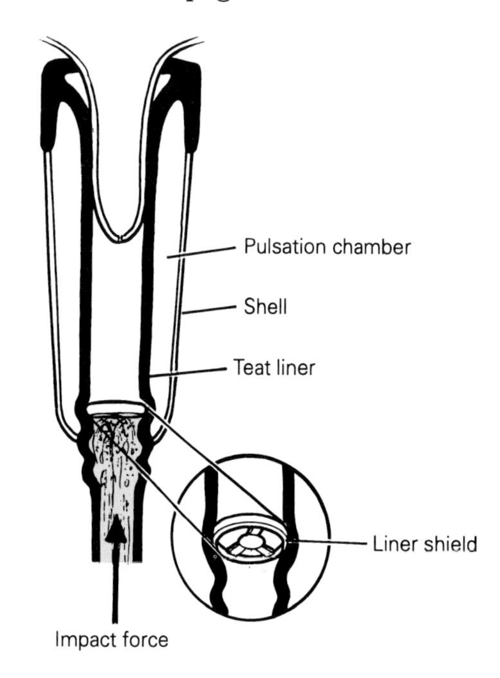

FIGURE 5.7 Liner shields help to reduce the effect of impact forces.

Experimental work has shown that liner shields have helped to reduce the new infection rate by up to 12%. However, the siting of shields is critical. If they are situated too high up the liner they may prevent total liner collapse. There will then be an incomplete massage phase and teat-end damage may result.

In normal milking, air enters the cluster through the air bleed hole to assist liner opening. However, when the air bleed is absent, the liners do not open and close fully and they then maintain close contact with the teat throughout milking. This is called **hydraulic milking** and results in the teats being bathed in milk throughout milking. The effect of hydraulic milking on the new infection rate is unknown at the present time.

MAINTENANCE AND MACHINE TESTING

Like any other piece of equipment, the milking machine needs to be maintained correctly so that it operates at maximum efficiency. An inefficient machine may at best slow down milking or at worst increase the amount of mastitis.

It is important that all the people who use the plant know what checks need to be carried out. A machine that is not operating to its full potential will still remove milk from the udder although it may well cause a predisposition to mastitis. Problems with milking machines tend to be gradual in onset and so checks need to be made regularly.

Some checks should be carried out daily by milkers, others at regular intervals by the manager or owner and others using specialist testing equipment on a routine basis. Most manufacturers now have a list of checks, together with a timetable of when these should take place.

The milking machine is a highly complex piece of equipment. It is used twice or sometimes three times every day of every year. Compare it to a motor car: car manufacturers recommend servicing every 5,000 or 10,000 miles. Virtually everyone gets their car serviced at the correct intervals: if not, the performance of the car starts to fall off and breakdown is more likely.

The milking machine is no different. The average milking machine runs for two and a half hours every milking. This is equivalent to over 1,800 hours of use each year. If you compare the milking machine to a car travelling at 40 miles per hour over a year, the plant would have done the same amount of work as a car travelling 72,000 miles!

Daily Checks by Milkers

The milkers are the first people who may feel that the machine is not operating correctly. Milking time may be increased or units may fall off for no apparent reason. Perished rubberware or parts such as worn valves etc. should be replaced as and when they are identified. If the milker is unable to identify and correct faults then the farm manager or machine dealer should be called in to sort out the problem.

Weekly Checks by the Manager or Owner

It would be unfair to leave all the responsibility for the milking machine to the milker. The manager or owner should check the plant regularly for any possible faults. Rubberware should be checked, liner condition inspected, the air filter on the regulator looked at and the oil level and belt tension of the vacuum pump etc. checked.

Routine Specialist Testing

It is important that milking machines are tested by a qualified technician or advisor on a regular basis. Most farmers have their plant tested once a year and some twice a year. There are some dairy farmers who have never had their plant tested since the day it was installed – sometimes a gap of up to twenty years! In the United Kingdom specialist testing of parlours should be carried out to British Standards, BS5545. These standards lay down the testing procedure and some acceptable performance criteria.

All milking machines should be tested every six months. Plants that milk three times a day should be tested more frequently. In Arizona, where three times a day milking is the norm, all milking plants are checked by the dairy company every month to ensure that they are operating at maximum efficiency.

There are two types of test that can be carried out: a static and a dynamic test.

The **static test** is carried out between milkings when no cows are being milked. This is equivalent to an MOT or mechanical inspection of a car. It is very useful in identifying certain problems, but it does have limitations. The **dynamic test** is carried out during milking and this is equivalent to a road test of a car. During the dynamic test the machine is tested under 'load' to see what, if any, problems are present.

Static Test

The static test includes the following:

Vacuum levels in the plant

Vacuum levels are checked at various locations throughout the plant to ensure that there is no significant loss of vacuum between the pump and the teat end. A drop in vacuum level would indicate that air is leaking into the system. The accuracy of the vacuum gauge is also checked.

Vacuum reserve

Vacuum reserve has already been described as being the production of vacuum over and above that needed to operate the plant. Adequate vacuum reserve is needed to ensure that stability of pressure is maintained in the plant throughout milking.

In England, the British Standards Institution (BSI) has made recommendations for vacuum reserve. It must be remembered that these are **minimum** recommendations and ideally new plants should exceed these levels. This will ensure that plants will still be able to operate to British standards as they get older and when their performance starts to become less efficient.

There is considerable difference between the recommendations for vacuum reserve in the United Kingdom and the United States as shown in Table 5.3. Many farmers feel that even British standards are overgenerous. However, an American plant with 16 milking units would require almost ten times the British recommended level of vacuum reserve in order to meet the standards in the United States!

Table 5.3 A comparison of the effective vacuum reserve (l/min) standards between the United Kingdom and the United States (*20*).

No. of milking units		*4*	*8*	*12*	*16*	*20*
Recommended vacuum reserve	UK	200	300	370	410	450
	USA	1000	2000	3000	4000	5000

Systems with a low vacuum reserve will have difficulty in maintaining stable vacuum levels during milking. This may result in an increased number of irregular vacuum fluctuations which may affect the incidence of mastitis. Research has shown that there is a correlation between herds where the milking machine has inadequate vacuum reserve and high cell counts (see page 125).

If units or other items that need vacuum are added to the milking system without any increase in the size of vacuum pump, then the level of vacuum reserve will be reduced. This may affect the degree of vacuum fluctuation.

The actual level of vacuum reserve is to some extent of academic interest. The important question is whether the vacuum requirements for milking and cleaning are satisfied.

Regulator function

It is important that the regulator functions correctly so that a stable vacuum level can be maintained throughout milking. Regulators most commonly become blocked with dirt, so reducing the amount of air leaking into the system, but occasionally mechanisms become defective. Regulators should be inspected and cleaned every week.

Pulsation system

The pulsation performance should be measured at each individual milking unit. In systems with master pulsation there should be little difference between units. In systems where there are individual pulsators, the difference in performance can be quite considerable.

General condition of the plant, rubberware etc.

The plant should be examined for any perished rubberware, leaky valves etc. and its overall condition noted. Liner condition should be assessed and the frequency of change checked to ensure that they are replaced at the correct intervals.

Dynamic Test

The dynamic test is carried out during milking. Vacuum levels and fluctuations close to the teat end are recorded during the milkout of a few high-yielding cows at the furthest end of the milking system, i.e. at the greatest distance from the vacuum pump and receiver vessel. This is designed to test the plant under 'load'. The length of milking time together with yield are recorded. The vacuum recording may be measured where the long milk tube leaves the milking cluster, or in the short milk tube just below the liner shell as shown in Plate 5.13. Rear quarters are preferred as they yield more than the fore quarters and this places the system under further load. The level and type of vacuum fluctuation is recorded.

There are two types of vacuum fluctuation that may occur at the teat end: regular and irregular. **Regular vacuum fluctuation** tends to be constant throughout milking. Figure 5.8 shows some different types of fluctuation that may occur during milking. Regular fluctuations are caused by pulsation. **Irregular fluctuations** occur as a result of factors such as inadequate vacuum reserve, unit fall-off, liner slip etc. and occur intermittently throughout milking. Liner slips are dangerous as they may result in reverse milk flow leading to impact forces. **Impact forces** drive milk droplets up against or through the teat canal and are discussed on page 63.

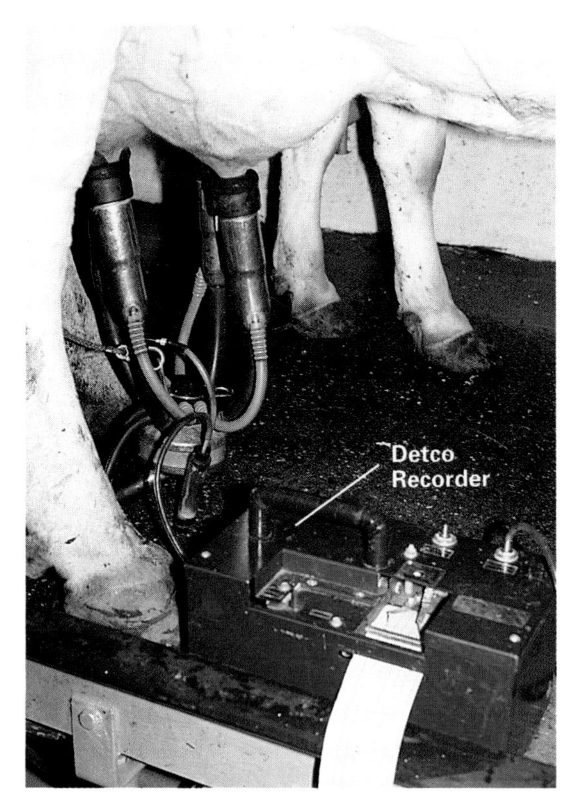

PLATE 5.13 A dynamic test recording the vacuum level and fluctuation at the top of the short milk tube. Note that the reading is taken at a rear quarter.

There are no fixed international standards for vacuum fluctuation. However, some American workers have suggested that levels over 10 kPa (3"Hg) are undesirable and may increase the risk of mastitis. **Any** irregular vacuum fluctuations are undesirable.

A continuous recording at the receiver vessel or the sanitary trap is also made over a ten minute period to check that there are no vacuum changes occurring in the plant. If there is any vacuum fluctuation here it is likely to be further exaggerated towards the teat end.

The behaviour of the cows should be noted during milking: are they comfortable and content, or are they edgy and uncomfortable?

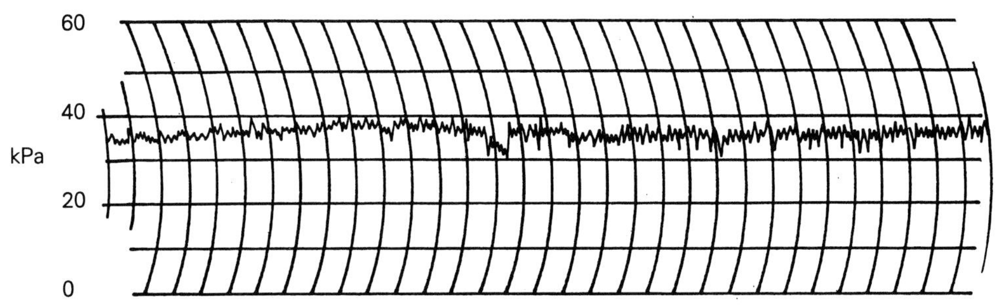

FIGURE 5.8a Good tracing showing little teat-end fluctuation.

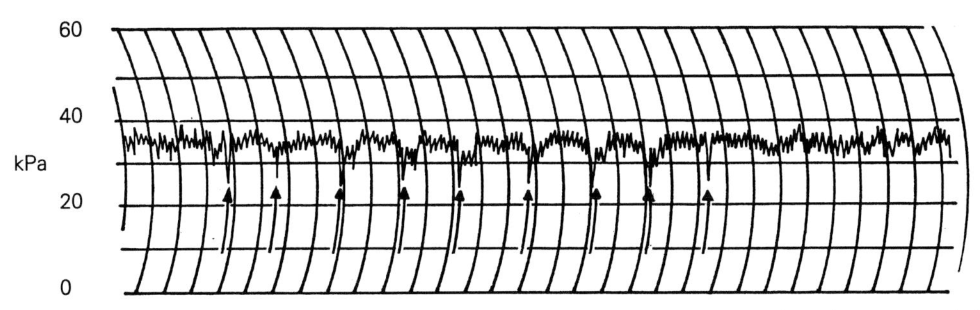

FIGURE 5.8b When other milking units are attached, the vacuum level at the teat-end drops as shown by the arrows. This suggests insufficient vacuum reserve or a problem with the regulator.

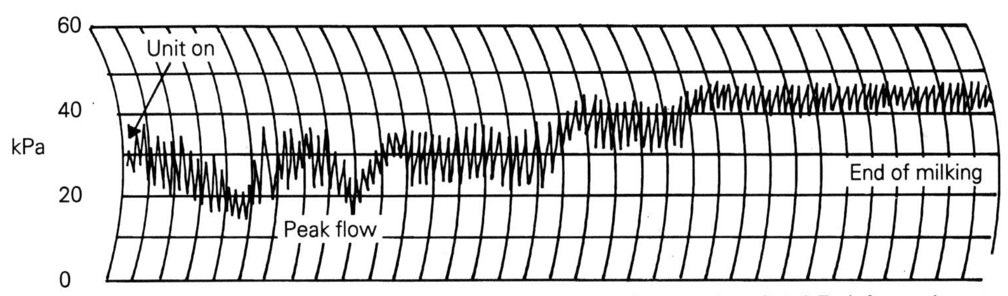

FIGURE 5.8c There is a high level of teat-end vacuum fluctuation (20 kPa) from the start of milking to peak flow, which reaches a more acceptable fluctuation (7 kPa) towards the end of milking. The actual vacuum level varies between 13 and 36 kPa at peak flow and between 40 and 50 kPa towards the end of milking.

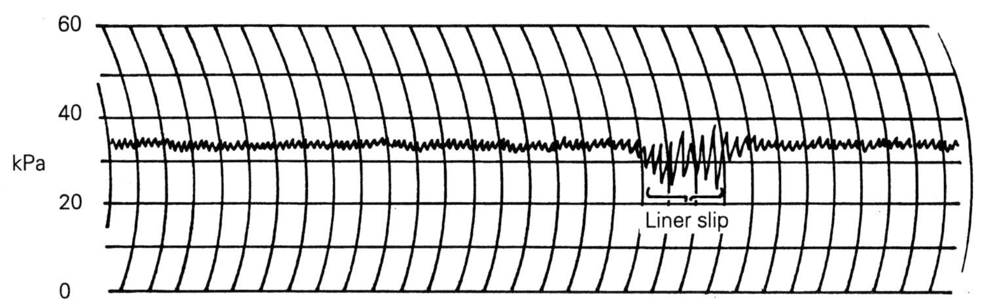

FIGURE 5.8d Teat-end vacuum level remains constant throughout except when there is a high level (17 kPa) of irregular fluctuation caused by liner slip. This may result in impact forces.

Occasionally problems may occur due to **stray voltage**. Stray voltage occurs if the parlour is not properly earthed and small amounts of electric current can pass through the metalwork. This causes restlessness in cows, especially paddling of the feet, and a resentment of milking. While this is uncommon in Europe, it is a well-known problem in North America and New Zealand. In the event of this being suspected, an electrician should be consulted.

It is essential that all the findings from the milking machine test are recorded and a report left on the farm. Test reports should **always** be discussed with both the milker and the owner and any faults identified related back to udder health.

The best machine test report is the one showing that no faults have been identified. In some cases when a test report has been left on the farm recommending immediate action, the message is ignored. This may occur because the significance of the fault has not been fully explained to the farmer and he sees no reason to take any action.

THE MILKING MACHINE AND ITS RELATIONSHIP TO MASTITIS

The milking machine can have an effect on the incidence of mastitis in one of four ways. It can:

- act as a fomite (i.e. a vector)
- damage the teat end
- increase bacterial colonisation at the teat end
- create impact forces

Acting as a Fomite

Disease organisms may be physically transmitted from the machine to cows, and in this way mastitis bacteria can be passed from cow to cow. This can occur through contaminated milk remaining on the liner between milkings (Plate 5.14). Research work has shown that following the milking of an infected cow,

PLATE 5.14 Infected milk remaining on the liner may spread infection to the next six to eight cows milked.

Staphylococcus aureus infection can be spread to the next six to eight cows milked through the contaminated liner. The risk of this occurring is increased if worn liners are used because bacteria are able to adhere more easily to their roughened surface.

Infection may also spread rapidly between quarters during milking if there is flooding of milk back up the short milk tubes. If this occurs milk from all quarters will mix allowing bacteria to contaminate uninfected teats. The amount of contamination will depend on how quickly milk is removed from the liner. Large capacity claws help limit this effect.

Damage to the Teat End

The importance of the teat canal mechanism in keeping the udder free from infection has been discussed on page 17. Milking cows through a faulty machine that damages the teat end will increase the risk of new infections.

There have been suggestions that overmilking and high vacuum levels lead to increased levels of teat damage. There is no evidence to show this to be the case

provided the machine is functioning correctly. This again highlights the importance of regular maintenance and testing.

Consider overmilking. About 60% of all milk produced comes from the rear quarters leaving 40% from the fore quarters which tend to milk out first. In effect, this means that the two front quarters are always overmilked. If overmilking contributed to any increase in mastitis, we would expect to find more cases of mastitis in the fore quarters and less in the rear. In fact the opposite is the case: 60% of all cases of mastitis occur in the rear quarters and only 40% in the fore quarters. This makes it unlikely that overmilking contributes to increased levels of mastitis.

Colonisation of the Teat Canal

Pulsation is important as it allows regular removal of excess keratin from the teat canal during milking. Keratin acts as a type of blotting paper mopping up any bacteria present and as on the skin, natural sloughing (removal) of superficial cells helps in the removal of bacteria.

If there are problems with pulsation and the excess keratin is not removed, then there will be a build-up of bacteria in the teat canal. This accumulation is due to reduced milk flow rates meaning that the excess keratin and bacteria are not 'stripped' away from the teat duct. These bacteria will be able to colonise the canal, and if they penetrate the udder, the risk of mastitis is increased.

Worn or an incorrect design of liner, small volume clawpieces, narrow bore pulsation tubes, and many other factors may affect pulsation. Problems with pulsation may also occur with short-barrelled liners as the short barrel may cause there to be insufficient space for the liner to collapse fully around the tip of the teat.

Impact Forces

Impact forces result in milk particles being propelled from the short milk tube or clawpiece, up against the teat end, as shown in Figure 5.9. They occur when

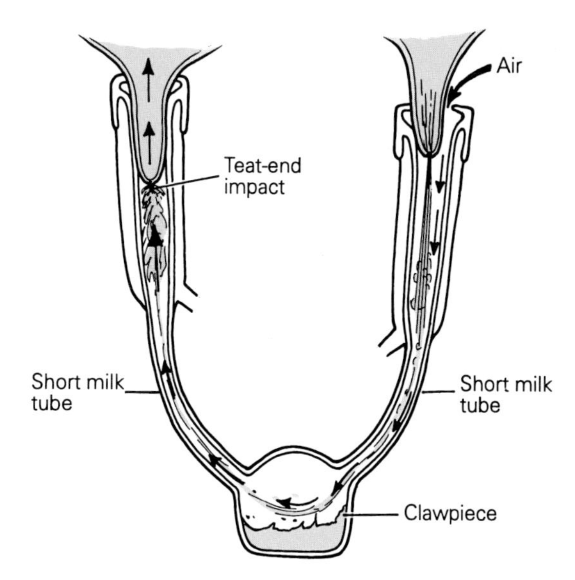

FIGURE 5.9 Teat-end impacts are caused when air enters between the teat and the liner, leading to an imbalance of pressure between the teat end and the clawpiece.

there is a pressure difference between the teat end and the cluster, often due to liner slip. This difference need only occur for milliseconds to create impacts. Milk may be driven at speeds of up to 40 miles per hour. This force is such that penetration of the canal, which is more open during milking (see page 18), may occur. If the milk is contaminated with mastitis bacteria then infection may follow. Pulling the milking unit off the cow without first shutting off the vacuum, machine stripping or liner slip (see Plate 5.15) may all result in impact forces.

Liner slips may occur due to one or more of the following reasons:

- milking cows with wet teats
- worn or perished liners

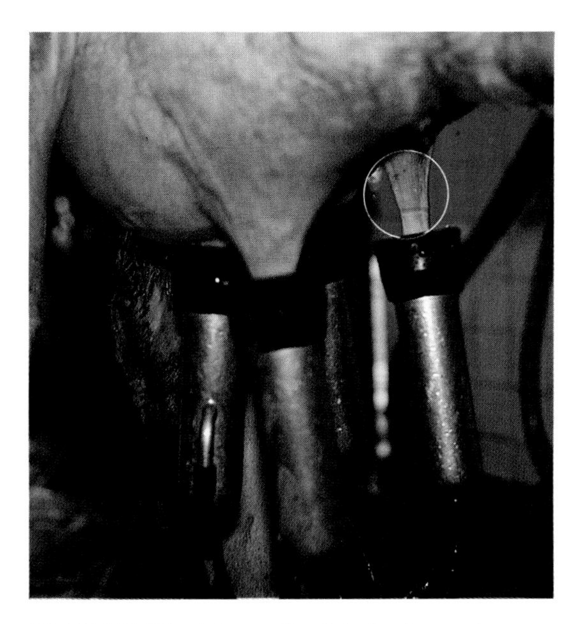

PLATE 5.15 Liner slip (circled) creates impact forces that drive milk against the teat end.

- low vacuum levels
- poor liner design
- heavy cluster weight
- cows with very small or very large teats
- high vacuum fluctuation during milking

The majority of liner slips result in 'squawking' of air as it enters through the top of the liner. As most liner slips occur towards the end of milking, they pose two dangers. Firstly, there is little resistance at the teat end because the teat canal is at its most open phase (see page 18). Secondly, if milk does penetrate the teat canal, because there is little milk left to be removed, it is more likely that bacteria will remain in the udder until the next milking. This will allow time for multiplication and possibly for mastitis to develop.

There has been a considerable reduction in the incidence of impact forces over the past 20 or so years. This is due to many improvements in milking machines including:

- larger pump capacity
- improved vacuum stability
- larger volume clawpieces
- larger diameter short milk tube
- larger air bleed holes
- liner shields

SIMPLE MACHINE CHECKS THAT CAN BE CARRIED OUT WITHOUT TESTING EQUIPMENT

There are a variety of simple tests that can be carried out to help identify possible machine problems without any of the sophisticated testing equipment used by specialists. These are described in the following section and are designed to help identify possible problems in the milking plant. They are not intended to replace routine machine testing which remains an essential part of any mastitis control programme. Specialist testing procedures will not be discussed.

Vacuum Level

- check that the vacuum gauge reads zero before the machine is switched on. Vacuum gauges are frequently faulty.
- when the machine is turned on, watch the needle rise. Most plants should reach operating vacuum level within about 10 seconds. If It takes a long time to reach the operating level then check that no valves have been left open – they may be leaking air into the plant.
- tap the gauge to check it is not sticking.
- what is the operating vacuum level according to the gauge? Check this against the last plant test report.

Vacuum Reserve

- set up the plant for milking. For every five milking units, open one so that it sucks air into the system. If the vacuum level falls by more than 2 kPa (0.6"Hg), it suggests that there is insufficient vacuum reserve.

- stand in the pit so that you have a clear view of the vacuum gauge. Leak air into the system through one milking unit for 5 seconds. Measure how much the vacuum level has dropped. Close the unit and record the time it takes for vacuum to return to the normal operating level. This is called the **vacuum recovery time**. It should never exceed 3 seconds. In an efficient plant the vacuum level should not drop at all during the vacuum recovery test.
- now repeat the test, leaking air in through two milking units for 5 seconds. How far did the level drop and how long did it take to return to normal this time?

In any system, leaking air in through one unit is equivalent to putting a unit on a cow. There should be no change in vacuum level. If the vacuum drops, it indicates that there is a problem – possibly inadequate vacuum reserve or perhaps the regulator is faulty or dirty. When two units leak air into the system there may be a small drop in vacuum. This should not exceed 10% of the plant vacuum level. So, if a plant operating at 45 kPa with one milker drops to 41 kPa when two units leak air into the system it is acceptable. For large plants with more than one milker, you should be able to leak air into the plant through a number of units equal to the number of milkers plus one, with little effect on vacuum, i.e. leaking air in through three units if there are two milkers.

Regulator Function

- first check that the regulator is clean.
- stand close to the regulator and listen. Can you hear air being admitted? The regulator should be continuously sucking atmospheric air into the system as there should always be surplus vacuum available throughout milking. What happens when units are put on, feeders cut in etc.? If the regulator stops leaking in air it indicates that the plant is unable to maintain a stable vacuum level.
- leak air into the system so that the vacuum level drops by 2 or 3 kPa. Listen to the regulator. If it is working correctly, it should **not** admit air as the plant is attempting to raise the working vacuum level back to normal. If it is sucking in air, then the regulator is faulty.
- watch the gauge rise to normal after lowering the plant vacuum level. If the gauge 'overshoots' and rises to above the normal operating level and then settles down, it suggests that it is either dirty or faulty.

Pulsation System

- first check if the pulsation is master or individual, single or dual.
- listen to each pulsator and check that it is working.
- place a finger into a liner of each cluster and check that it is moving. In the case of dual pulsation you will need to check two liners with alternate movement. If no movement is felt, it suggests that there may be defective pulsation in this unit.
- measure the pulsation rate. Check this result against the last machine test report.
- check the condition of the short and long pulsation tubes. A hole in a pulsation tube will affect liner movement. The liner may only partially open, or may not open at all as it will be unable to create a full vacuum in the pulsation chamber. This will depend on the size and the location of the hole.

It is easy to demonstrate the effect of inadequate pulsation to the milker: place bungs in three liners and ask the milker to put their thumb inside the open liner. Turn on the vacuum supply as for milking and kink the long pulsation tube. This will mimic continuous milkout without pulsation and will stop blood circulation around the teat or thumb.

Watch the effect on the milker and see their red and swollen thumb afterwards as shown in Plate 5.16.

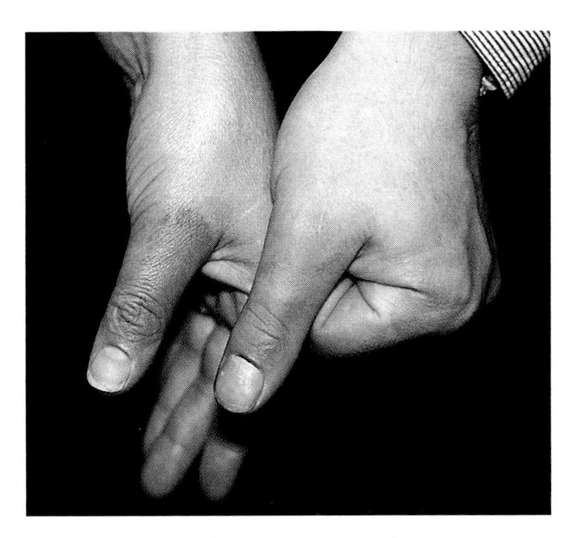

PLATE 5.16 If there is no pulsation, blood circulation stops as shown in the 'thumb test'.

Liners and Rubberware

- while checking the pulsator action, feel the inside of the liner. Is it soft and smooth, or rough and cracked?
- if you have a pencil torch, shine it inside the liner and have a look.
- how often are the liners changed?
- how frequently should they be changed?
- take a liner out of its shell: is it collapsed or round?
- split the liner lengthways and look at its condition. (Always make sure that you have a replacement liner before you do this or milking will be difficult!) Is the liner clean?
- check the condition of the rest of the rubberware for cracks, holes or splits.

Other Checks

- check the air bleed hole on the cluster. If it is blocked, milk will not flow away from the cluster easily. If milk runs back out from the cluster when it is removed from the cow, it indicates that the air bleed hole is blocked.
- when was the machine last tested?
- what tests were carried out?
- have all the problems identified at previous visits been corrected?
- was a report left and were the results discussed fully?
- who usually tests the plant?

Checks to be Carried Out During Milking

- check that the units sit on the udder comfortably.
- is there any evidence of liner slip?
- do units fall off cows for no apparent reason?
- are cows happy during milking or kicking and restless?
- check the function of the ACRs.
- are cows under or overmilked?
- check that the vacuum gauge remains static during milking.
- look for evidence of teat damage such as small haemorrhages, sphincter eversion, distortion and cyanosis of the teat skin (see page 180) as soon as the cluster is removed from the cow.

WASH-UP ROUTINES

An efficient parlour wash-up routine will remove milk residues and bacteria from the plant. This will maintain milk quality, improve the appearance of the parlour and prolong the life of milking equipment. Problems with the wash-up routine will result in milk residue and bacterial build-up within the system. This will increase the TBC (see page 135). The milking system is washed through the physical action of cleaning solutions assisted by temperature and chemicals. No matter what system is used, the machine will not be adequately cleaned unless wash-up solutions come into contact with **all** soiled parts of the plant. All too often poor cleaning is due to poor circulation of solutions, blocked jetters, low temperature or an inadequate volume of wash solution.

The following will be required irrespective of the type of milking system:

- a supply of potable water (water free from faecal contamination).
- an efficient water heater.
- a thermometer.
- storage for chemicals.
- protective clothing.
- For circulation cleaning, one if not two wash troughs.

British Standards require a **minimum** of 18 litres (4 gallons) of hot water per milking unit for circulation cleaning or acid boiling wash. Less than this and TBCs may increase. Remember that some hot water may be used for other purposes such as feeding calves, etc. Plants with large bore pipelines may need more than 18 litres boiler capacity per unit.

The milking system should be cleaned immediately after milking while the plant is warm and before milk deposits start to form on pipes. Two forms of cleaning are used: circulation cleaning and acid boiling wash (ABW). Circulation cleaning is the most common method in the United Kingdom.

The milking system is designed to produce minimal turbulence of milk during milking because excess turbulence may lead to 'buttering' of milk. However, during the wash-up routine maximum turbulence is required to make sure that all internal surfaces of the plant are cleaned. In some parlours more vacuum reserve will be needed to draw wash solutions through the plant than is needed to provide stable vacuum during milking.

Some plants are fitted with **air injectors** that admit 'slugs' of air to increase the turbulence by bubbling and swirling water all around the pipes. Air injectors are of increasing importance in large bore systems so that the entire surface of each line can be physically cleaned. Plate 5.17 shows an air injector sited at the junction of the wash line and the milk transfer line.

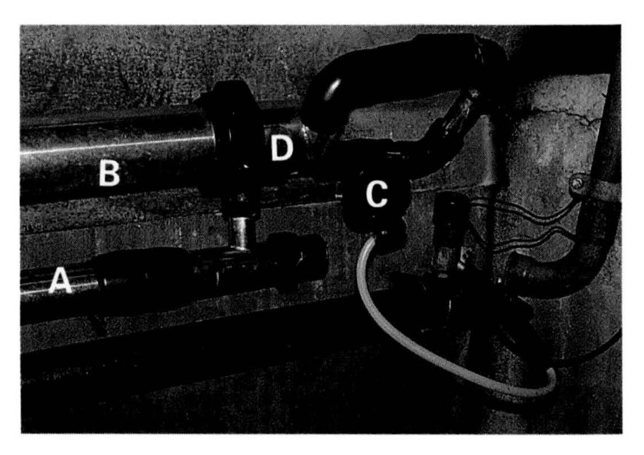

PLATE 5.17 Water coming through the wash line (A) flows into the milk transfer line or long milk line (B), and then flows back to the receiver vessel. The air injector (C) releases pulses of air into the milk transfer line (B), which causes a swirling action. The dead end (D) is undesirable as it may be difficult to clean.

It is essential that all dairy chemicals are stored safely and used correctly. Protective clothing (goggles, gloves and aprons) should be worn when handling chemicals to avoid accidental injury.

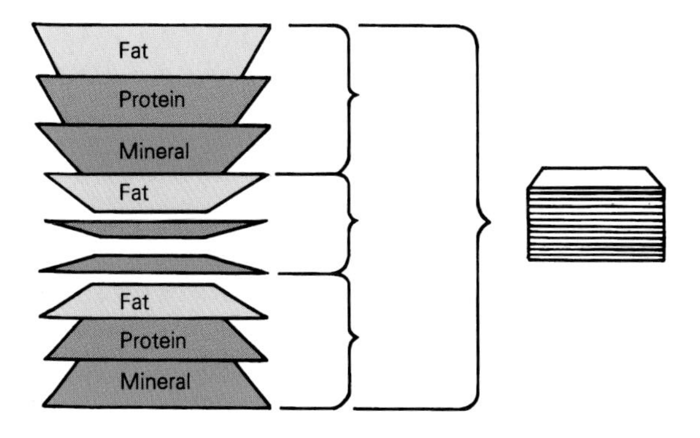

FIGURE 5.10 If problems occur with the wash-up routine, milk films will build up in the plant. This diagram shows how milk films develop.

Chemicals should not be stored in the same room as the bulk tank to avoid any risk of milk taint occurring.

If problems occur with the wash-up routine, then milk films can build up in the plant, as shown in Figure 5.10. These films provide nutrients for bacteria which can then multiply and increase the TBC (see page 135).

There are two types of milk soil: **organic** and **inorganic**. Organic soils are composed of milk fat (butterfat), protein and sugar. If these residues are left in the plant they will harden as they dry. Inorganic soils result from mineral deposits such as calcium, magnesium and iron and are often referred to as 'milkstone'.

Butterfat

The temperature and pH of the cleaning solution must be right in order to ensure that all butterfat is removed. Butterfat starts to solidify at temperatures under 35°C. This is an important consideration for the rinse cycle of circulation cleaning.

Alkaline detergents are used to remove butterfat and must be capable of emulsifying (breaking down) the fat globules so that they can be removed from the system. If butterfat solutions build up in the plant they trap other forms of milk soil and also have a detrimental effect on rubber components.

Protein

Protein films are hard to see. They adhere strongly to pipes and are difficult to remove. Alkaline detergents with chlorine added (chlorinated) can break down protein so that it can be removed from the plant. However, water at high temperatures can bind protein deposits onto milk pipes.

Minerals

Calcium, magnesium and iron may cause problems if allowed to precipitate, particularly in hard water areas where milkstone can easily build up on the pipes. This is seen as a chalky film. Acid solutions are used to remove and prevent the accumulation of mineral deposits and most dairy farmers run an acid wash (milkstone remover) through the plant on a regular basis. These washes usually contain phosphoric acid.

Bulk Tanks

The bulk tank must be cleaned every time milk is emptied from it. Most tanks are now cleaned by automatic washers that rely on chemicals and jetters to complete the process. All internal parts of the tank that can come into contact with milk must be cleaned and disinfected. It is essential that the milker checks that this process has been carried out efficiently before milking.

While automatic washers do a good job, there is the risk that because they are automatic, things can go wrong. The milker rarely looks at the tank to check on cleanliness. Occasionally problems do occur such as when a wash jetter becomes blocked and so part of the tank is not cleaned. This causes a build-up of milk film in which psychotrophic bacteria can multiply, resulting in increased TBCs. Psychotrophs are bacteria that thrive under refrigerated conditions.

Airlines

Occasionally milk does enter the airlines. For example, this can occur if there is a split liner. In this case milk will be sucked up through the pulsation chamber, into the pulsation tubes and into the main pulsation line. If milk does enter airlines, then these need to be cleaned out. All airlines should be washed twice a year.

CIRCULATION CLEANING

This form of cleaning is divided into three cycles:

- rinse } remove soil from the plant
- wash }
- disinfection – removes residual bacteria from the cleaned plant

When milking is completed any milk in the receiver vessel and the milk pump should be drained. The milk pipe should be disconnected from the bulk tank. The external surfaces of clusters and milking units should be rinsed clean (ideally with warm water) and the plant set up for the wash-up routine. This consists primarily of attaching jetters to the clusters and then transferring vacuum to the wash lines so that wash water is drawn into the wash lines, through the cluster and back through the return wash line as shown in Figure 5.11.

Rinse Cycle

Warm water at 38–43°C should be rinsed through the milking system and run to waste until the water appears clear, immediately after milking. This will remove the majority of any residual milk left in the plant. Most dairy farmers, however, use a cold rinse. Under no circumstances should cold water be used as it will congeal butterfat onto glass and stainless steel fittings and cool down the plant before the hot wash. Energy in the form of hot water will then be required to heat up the pipes before the hot wash. In addition, minerals and sugars in milk are

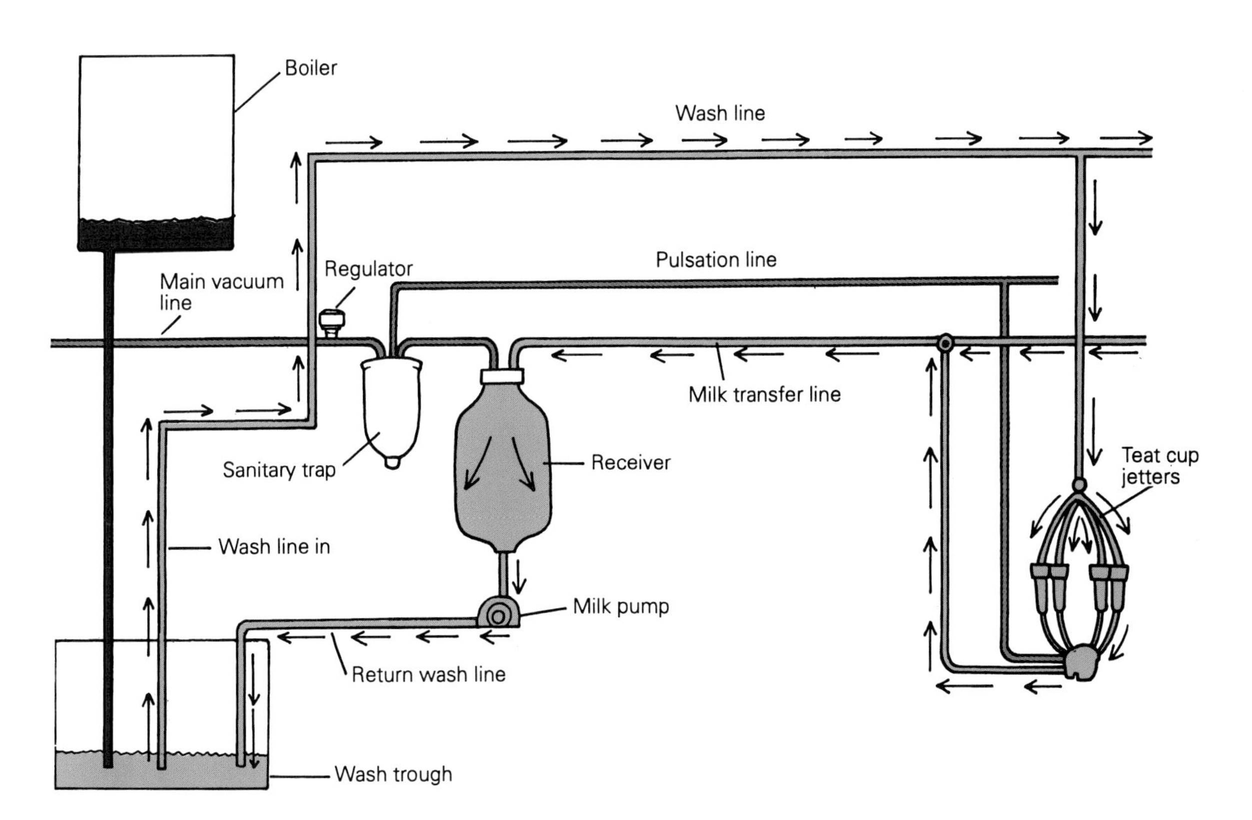

FIGURE 5.11 With circulation cleaning, water is drawn into the wash lines, through the cluster, back through the milk transfer line and into the wash trough where it is recirculated.

more easily dissolved in warm water.

After the rinse cycle, the wash line valves should be shut off to prevent large volumes of air being sucked into the system. This air may cool down the plant before the hot wash. It is advisable to insulate the milk and wash lines. This is essential where any of these pipes are exposed to the outside environment. Not only will it prevent excessive cooling during the wash-up process, but it will also prevent milk freezing during winter milking. An efficient rinse cycle should remove between 90 and 95% of all milk residue in the system and all the milk sugars. The remaining residues are removed during the wash cycle by chemical action.

Wash Cycle

In circulation cleaning, the wash cycle relies on an alkaline detergent solution to remove butterfat. Chlorine is normally incorporated into the detergent to remove protein, but in this form the chlorine has no disinfectant property. Wash-up solutions are very sensitive to temperature. In general their cleaning power doubles for each 10°C increase up to a maximum of 71°C. Once they exceed this temperature they tend to become unstable, vapourise and hence become less effective. There are, however, a few detergent solutions designed to be circulated in cold water, so always follow the manufacturer's instructions.

An adequate supply of hot water must be available. Maintaining the correct temperature of water in the boiler is essential. It is also important that the boiler has a large bore tap so that the wash trough can be filled rapidly without heat loss.

The water temperature should be checked regularly against the boiler gauge to ensure that the thermostat and heater element are functioning correctly. Sometimes the boiler gauge becomes faulty or the heating element becomes caked with mineral deposit. This is especially common in hard water areas. The best way to check on boiler efficiency is to fill a trough with hot water and measure the temperature in the filled trough. This is the temperature that counts. The first water that runs out of the boiler will always be the hottest.

Detergent solutions must be used correctly and to do this the volume of wash water must be known. Check the manufacturer's instructions on how much detergent to use.

Hot water is run into the plant and as it travels around it heats up the pipes and vessels. Only then should the correct amount of detergent solution be mixed into the circulating hot water. It should be circulated at 60–70°C for 5–8 minutes, or in accordance with the manufacturer's recommendation. If the solution is too weak it will be ineffective. If it is too strong then it will be wasteful and may even corrode the stainless steel or rubberware in the plant.

The milker should not clean the outside of the recorder jars with cold water during the rinse or wash cycle as this will cool down the temperature of the jars and also the circulating solutions.

If the rinse cycle has not been effective a large amount of milk residue may be left in the system. This milk will inactivate some of the detergent which will decrease the effectiveness of the wash-up routine.

Ideally a thermometer should be fitted to the return wash pipe to check that the wash solutions are being circulated at the correct temperature in accordance with the manufacturer's recommendations.

If solutions are circulated for long periods of time, their temperature drops and then protein may be deposited back onto the pipes. Milkers have been known to let the hot wash cycle run while they go and have breakfast!

At the end of the circulation wash cycle, the milking plant should be clean and free from any milk soil.

Disinfection Cycle

The disinfectant rinse reduces the number of bacteria in the plant and helps maintain milk quality. Sodium hypochlorite is the most commonly used disinfectant, at a strength of 50 ppm (parts per million). This solution may be circulated and then dumped to waste at the end of the cycle or it may be left in the plant until the next milking. It will then have to be drained from the system before milking can take place.

If you examine the internal surfaces of some rubber components in the plant and a black deposit marks your finger, as shown in Plate 5.18, this indicates rubber damage caused by too high a level of hypochlorite. The same effect may be seen in the black colour of the disinfection rinse shown in Plate 5.19. Remember that chlorine compounds reduce the life of all rubberware and liners.

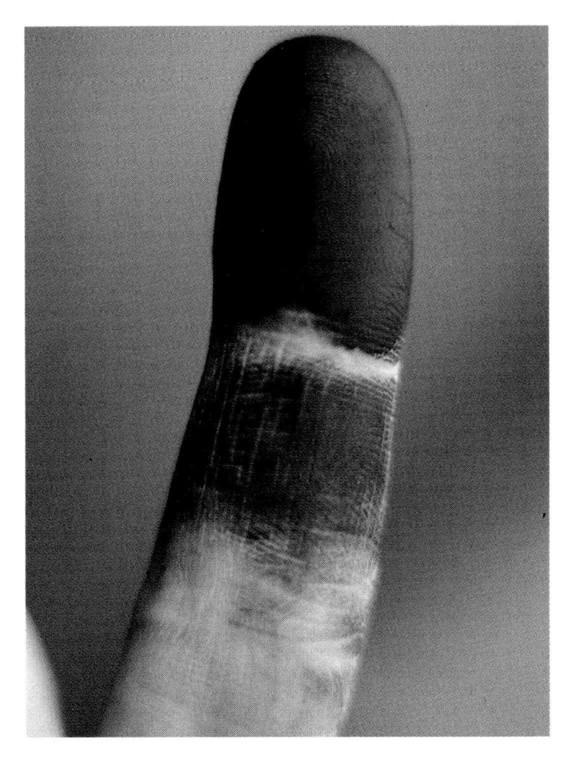

PLATE 5.18 Black deposits from rubber parts indicate high levels of hypochlorite, resulting in rubber corrosion.

Summary of Common Problems Associated with Circulation Cleaning

- water is not hot enough – this will make the detergent solutions less effective as they are temperature sensitive.
- inadequate volumes of wash – water may not come into contact with all internal surfaces. This may result in some areas of the system not being cleaned, especially the top part of the milk lines.
- rinsing the plant with cold water after milking – this will cool down the warm plant and congeal butterfat. The hot wash solutions will then have to heat up the plant from cold, and the detergents will have to remove the butterfat deposits.
- incorrect strength of detergents used – too little is ineffective while too much is costly and corrosive.
- wash cycle continued in excess of the recommended time – the solutions will cool down and may re-deposit material back onto the internal surfaces.

PLATE 5.19 High levels of hypochlorite in the disinfection solution strip away rubber.

- build-up of deposits in dead-end areas which are difficult to clean, as shown in Figure 5.12.

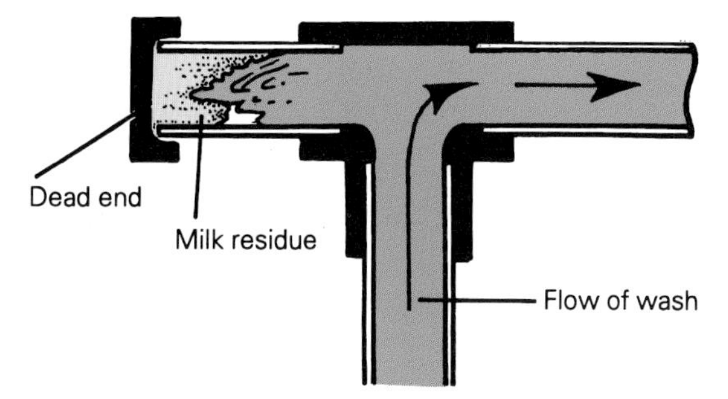

FIGURE 5.12 Deposits frequently build up in dead-end areas as they are difficult to clean.

- insufficient turbulence or flow of wash solutions – cleaning may be ineffective and deposits can accumulate on liners (see Plate 5.20). Build-up of milk soil occurred on the milk transfer line in this plant due to inadequate flow of wash solutions.

PLATE 5.20 If there is insufficient turbulence or flow of wash solutions, cleaning may be ineffective as shown on this pipeline.

- blocked wash-up jetters – this may result in one liner or a complete milking unit not being washed. The effect will depend on where the jetter is blocked.
- faulty air injectors – these will not create the physical turbulence that is needed to clean large bore lines.

ACID BOILING WASH (ABW)

ABW relies entirely on heat for disinfection. Large amounts, 18 litres (4 gallons) of boiling water over 96°C are needed for each milking unit. This water is run directly from the boiler around the plant to waste, as shown in Figure 5.13. It takes the same path as for circulation cleaning, the only difference being that the solutions are run around the plant to waste rather than circulated. All parts of the plant to be cleaned must reach and maintain a temperature of 77°C for the whole cycle, which lasts between 5 and 6 minutes. In the first few minutes of cleaning, a solution of dilute nitric or sulphanilic acid is run in the boiling water to prevent deposits building up on any of the surfaces. See Plate 5.21.

The plant must be capable of withstanding high temperatures and acids. There should be no dead-ends and

PLATE 5.21 Acid boiling wash (ABW) relies on heat and acid for cleaning. Acid is put into the reservoir. Boiling water is then run through the system for about 90 seconds to heat up the plant. The acid is then released and mixes with the boiling water to run around the plant to waste. The arrows show the acid running out of the reservoir and joining the flow of water.

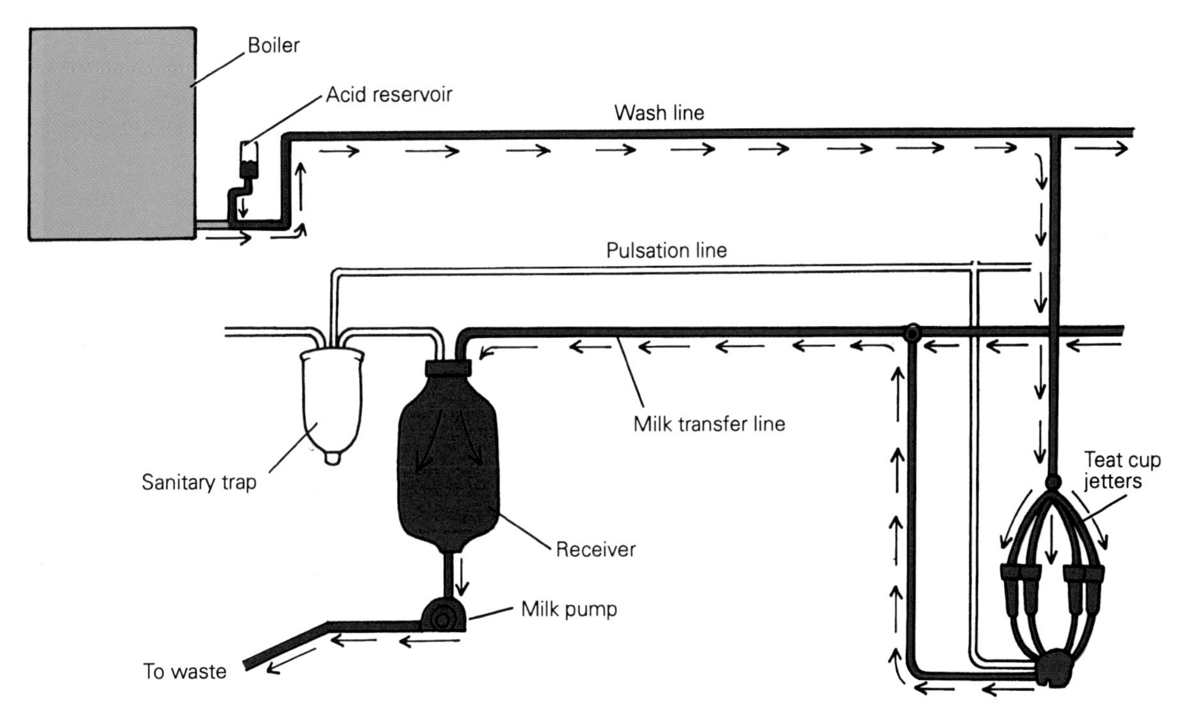

FIGURE 5.13 The route of wash solutions with acid boiling wash. The boiling water and acid are run through the plant to waste.

the whole system should be as compact as possible to avoid excessive heat loss. This form of cleaning is not very popular in England as problems occur if the water temperature or volume is too low. ABW does save on dairy detergents and is a faster form of washing compared to circulation cleaning. However, it does require the boiler to heat the water to a very high temperature.

MANUAL WASHING

Dump buckets and their clusters can either be washed as part of the wash-up routine or manually using chemicals and brushes in any system. While this process is labour intensive and time-consuming, excellent results can be achieved and in addition, the milker is able to visually check on the efficiency of the process. It is important that clusters are thoroughly cleaned and disinfected as they are often used to milk colostrum from freshly calved cows which are very prone to mastitis, see page 24.

Some farmers place clusters and other pieces of milking equipment in troughs with small volumes of detergent solutions, as shown in Plate 5.22, and expect good results. This has little effect as the solutions do not come into contact with all the internal surfaces. In addition, if this method of cleaning is used on clusters which milk mastitic cows, then this must increase the risk of spreading infection as liners may remain contaminated with mastitis organisms (see page 86).

IF A WASH-UP PROBLEM IS SUSPECTED

The efficiency of the wash-up routine can

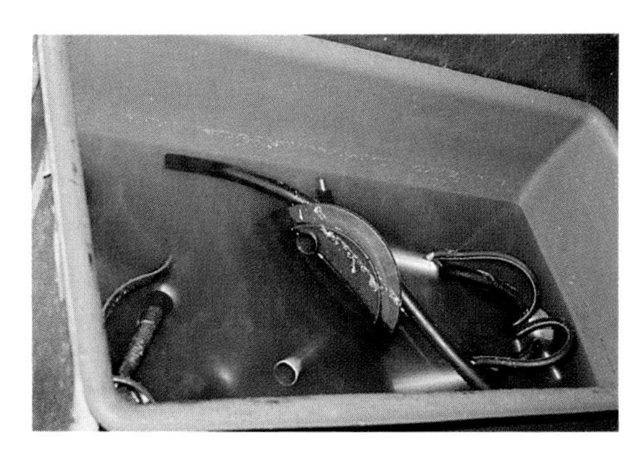

PLATE 5.22 Cleaning will be ineffective as there is insufficient detergent solution to come into contact with all parts of the cluster.

be evaluated through laboratory bacteriology of bulk milk as described on page 137. If a wash-up problem is suspected, the cause must be identified. Much can be gained by manually inspecting the system after the wash-up routine, for example by removing pipe ends and examining internal surfaces with a torch.

Look inside the following areas for any evidence of milk film or milk soil build-up:

- liners
- milk transfer lines (especially the top of the line)
- bungs and valves at the base of jars and lines
- ACR flow sensors
- receiver vessel
- dead-end areas
- milk pump
- bulk tank

COMMON FAULTS FOUND WITH MILKING MACHINES

Below is a list of commonly found problems and their consequences:

- small clawpieces (see Plate 5.23 middle claw) are likely to flood, impede rapid milk removal and increase cross-contamination between quarters (see page 62). The chances of irregular vacuum fluctuations and liner slip may also be increased, leading to a higher new infection rate.

PLATE 5.23 A variety of clawpieces.

- excess bends in pipes (Plate 5.24): these impede air and milk flow and may reduce the amount of vacuum available at the teat end, increasing the risk of irregular vacuum

PLATE 5.24 Excess bends in pipelines.

fluctuations in the plant, and leading to an increased infection rate.

- hole in the short air tube (Plate 5.25): atmospheric air is sucked in through the short pulsation tube. This is likely to prevent full vacuum levels being attained in the pulsation chamber, resulting in incomplete liner opening which will slow down milking in that quarter. In severe cases the liner may not open at all resulting in failure of that quarter to milk out.

PLATE 5.25 Hole in the short air tube.

- split liner (Plate 5.26): a split liner is unable to open and close normally. This has two effects. Firstly, it can result in incomplete milkout and massage, leading to teat damage, which may lead to an increased risk of new infections. Secondly, milk will get sucked into the pulsation chamber, up through the pulsation tubes and into the long pulsation line. This may affect pulsation itself.
- congestion of blood in the teat (Plate 5.27): this is not a common problem but can occur due to a variety of reasons including:
  - absence of pulsation, i.e. full vacuum constantly applied to the teats
  - incomplete or defective pulsation
  - excessive vacuum levels
  - poor liner design
  - using incompatible liners and shells

  Congestion of the teat causes the cow great discomfort and the milk let-down reflex is likely to be reduced. If teat damage occurs, this will increase the likelihood of mastitis.
- blocked regulator filter (Plate 5.28): the dirty regulator may be unable to

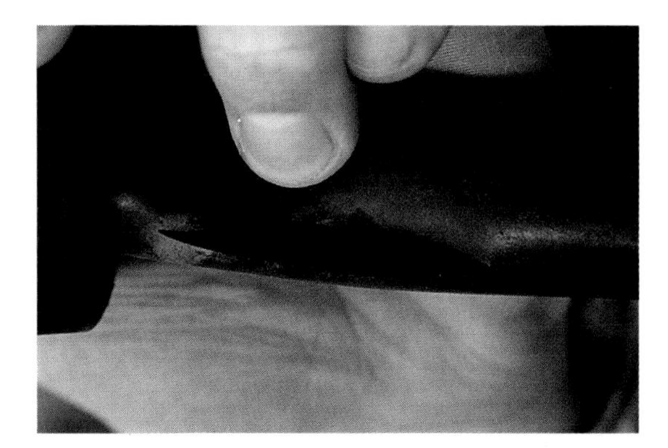

PLATE 5.26 Split liner.

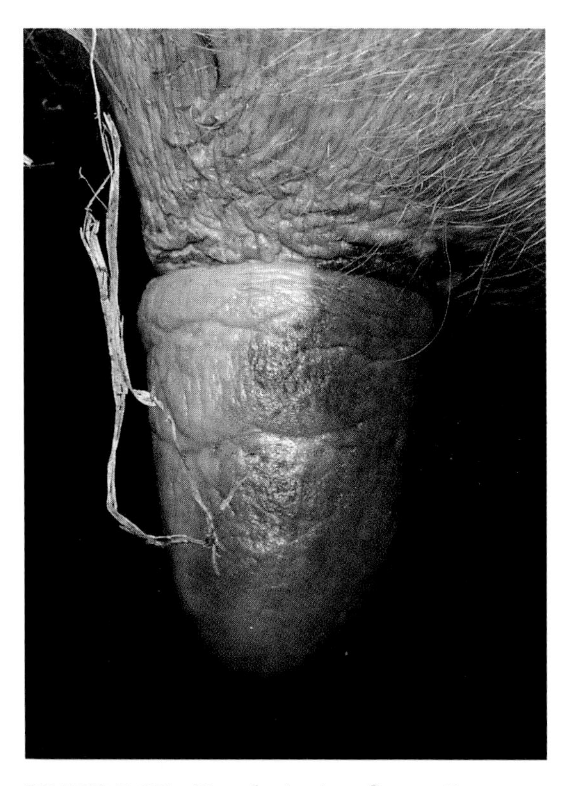

PLATE 5.27 Purple teats after unit removal. This is caused by congestion of blood in the teat.

respond rapidly to vacuum changes in the system resulting in poor vacuum stability. This can increase the likelihood of irregular vacuum fluctuations thereby increasing the risk of new infections.

PLATE 5.28 Dirty regulator. Note how inaccessible it is.

- multiple weight-controlled regulators (Plate 5.29): these act independently of one another. They all try to maintain stable vacuum in the system, however, as they respond slowly to pressure changes, it is likely that they may work against each other, resulting in poor vacuum stability.

PLATE 5.29 Multiple weight-controlled regulators.

- sanitary trap without a floating ball valve (Plate 5.30): the floating ball is designed to rise and cut off the vacuum supply if liquids build up in the jar. If the sanitary trap floods, milk will enter the vacuum and pulsation lines. This can affect the performance of the vacuum pump and pulsation, and contaminate the air (pulsation) lines.

PLATE 5.30 Sanitary trap without a floating ball valve. The value is necessary to prevent dirt etc. entering the pipelines.

CHAPTER SIX

The Milking Routine and its Effect on Mastitis

This chapter describes the various processes which constitute a good milking routine and discusses the way in which they can influence the incidence of mastitis.

A good milking routine will remove milk efficiently from the cow with minimal risk to udder health. It must include practices that limit the spread of contagious mastitis in the parlour. This results in quality milk production with low bacterial contamination. The milking routine should be designed to achieve these goals but at the same time it must be practical and labour efficient. The milker needs to understand the scientific reasoning for each step in the milking process in order to achieve these aims.

It is important that a consistent milking routine is practised in the herd. Cows love uniformity. They are also easily stressed and so rough handling or aggressive milkers are to be avoided. Cows readily become nervous and this can affect their milk let-down reflex. The conscientious dairyman will benefit from a good routine by reduced levels of mastitis, increased milk production and a rapid milkout.

The beneficial effects of hygiene at milking time are shown in Figure 6.1. It demonstrates how different hygiene practices can result in a reduction in clinical mastitis and also in the new infection rate. This is to be expected as contagious organisms are transferred between cows during the milking process.

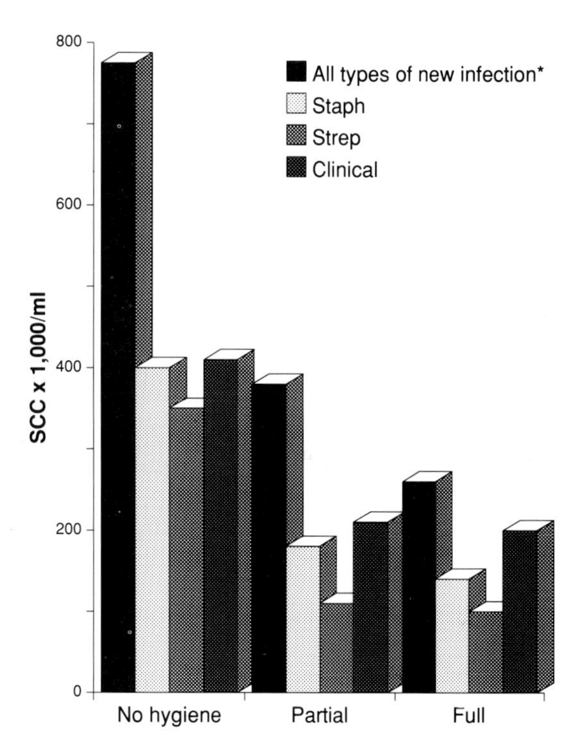

Hygiene	*None*	*Partial*	*Full*
Disinfectant udder wash	–	✓	✓
Individual cloths	–	✓	✓
Rubber gloves	–	✓	✓
Disinfectant hand dipping	–	✓	✓
Teat dipping	–	✓	✓
Pasteurised clusters	–	–	✓

* The number of new infections is higher than the number of clinical cases as many new infections become subclinical, 'hidden' or are eliminated from the udder without any outward signs of mastitis.

FIGURE 6.1 Effect of different hygiene regimes on new infection rate and clinical mastitis. (21)

There is only a slight reduction in infection between 'partial' and 'full' hygiene, indicating that the benefits of pasteurising clusters are limited in the United Kingdom (see page 86).

GLOVES

The milker can spread contagious mastitis as he handles each cow. It is extremely difficult to disinfect the rough surface of hands let alone keep them clean during milking (see Plate 6.1). For this reason it is advisable to wear rubber gloves.

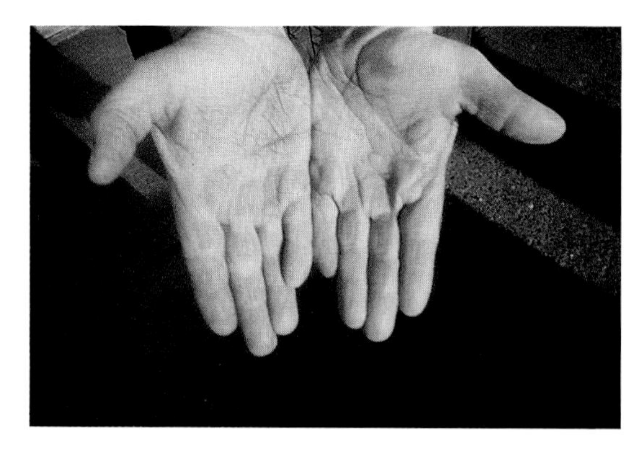

PLATE 6.1 Hands have rough surfaces that are difficult to clean compared to the smooth surfaces of rubber gloves.

Trial work in 1966 showed that half of all milkers' hands were infected with mastitis organisms even before milking had started! Contamination increased during milking so that by the end all milkers' hands were infected.

In another experiment two groups of milkers' hands were cleaned in different ways. The first group were washed with a disinfectant solution, and afterwards only 30% of hands remained contaminated. However, in the second group, whose hands were just washed in water, 95% remained infected.

It is essential that farmers have at least three pairs of gloves available. One can be worn during milking, while another is disinfected. The third pair may be needed if one of the others is damaged. Once gloves become worn or torn they must be discarded. Many milkers overcome this problem by wearing disposable gloves that are discarded after each milking. To be effective, gloves must of course be rinsed frequently in a disinfectant solution during the milking process.

The use of Rubber gloves are especially useful when dealing with *Staphylococcus aureus* or *Streptococcus agalactiae* infections. *Streptococcus agalactiae* has been isolated from milkers' hands up to 10 days after their last contact with infected animals. Indeed, some relief milkers are known to have introduced infection into clean herds in this way.

FOREMILKING

This is the practice of hand milking each teat before unit attachment. It is recommended for three reasons:

- aids detection of mastitis
- contributes to the milk let-down reflex
- flushes out the milk from the teat canal. This will remove most bacteria that have entered the teat since the previous milking, and hence will help to reduce the TBC.

Early detection of mastitis allows prompt treatment of clinical cases. This not only results in higher cure rates, but more importantly, it reduces the risk of spreading infection to the rest of the herd. It also stops mastitic milk from entering the bulk tank and this helps to avoid high TBC and cell counts. In the case of *Streptococcus agalactiae* and *Streptococcus uberis* infections, up to 100,000,000 organisms per ml of milk can be shed from an infected quarter. This can account for the fluctuating TBC levels that are frequently found in herds with an *Streptococcus agalactiae* or *Streptococcus uberis* problem (see page 134).

Frequently there are clots in the first two or three strippings from cows but the remainder of the milk appears normal. This is probably a response to bacteria in the teat sinus but not in the udder itself. In such cases only the foremilk needs to be discarded.

Foremilking for mastitis detection is time-consuming. In addition, there are possible disadvantages. For example, if you have a mastitis incidence of 45 cases per 100 cows per year, nearly 5,500 teats will have to be foremilked to detect one case of mastitis when milking twice a day. This increases to over 8,000 teats when milking three times daily! Some feel that there is a greater risk of spreading infection from cow to cow as the milker's hands or gloves become contaminated with mastitis organisms and consider that this outweighs the risk associated with a failure to detect mastitis at an early stage. This is particularly the case in herds with a low incidence of mastitis. Many herds now do not foremilk as a routine.

If practised, foremilking should be carried out before the teats are washed. Any milk that contaminates the milker's hands will then be washed off the teats before infection can spread to the next cow. Milk should be stripped onto the parlour floor rather than into a strip cup (see Plate 6.2) as strip cups tend to become reservoirs of infection rather than acting as an aid to detection. A black tile built into the parlour floor under each cow's udder allows easier examination of milk.

TEAT PREPARATION

One Somerset dairy farmer has a notice up in his dairy for his milkers, 'If the cows' teats are not clean enough to put in your mouth, then they are not clean enough to put the cluster on'. This sums up pre-milking teat preparation perfectly.

Good teat preparation is essential for clean milk production. It also helps to reduce the risk of environmental mastitis. The goal in teat preparation is to ensure that teats are clean and dry before the milking units are attached (see Plate 6.3). The best way to ensure that

PLATE 6.2 Milk should be stripped onto the parlour floor rather than into a strip cup. Ideally the milker should wear gloves.

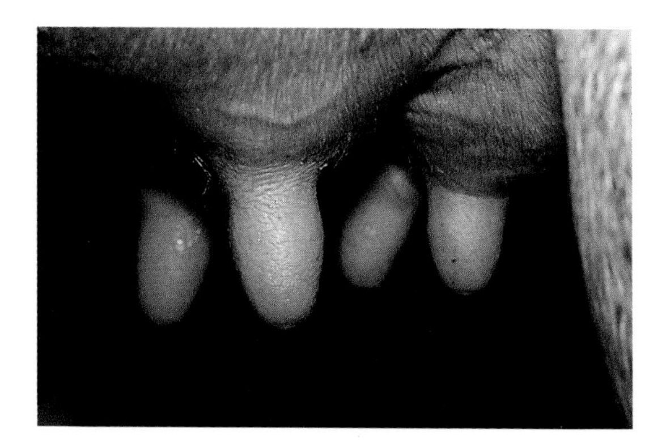

PLATE 6.3 Ensure that teats are clean and dry before the milking units are attached.

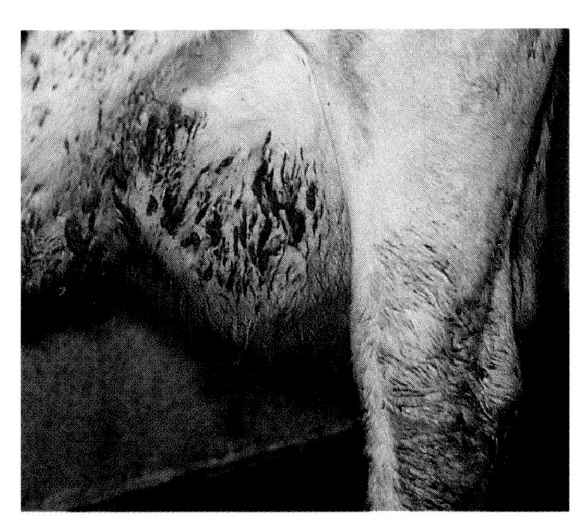

PLATE 6.4 Hairy udders are likely to trap dirt and hence increase the risk of environmental mastitis.

PLATE 6.5 Visibly clean teats can be dry wiped with a paper towel.

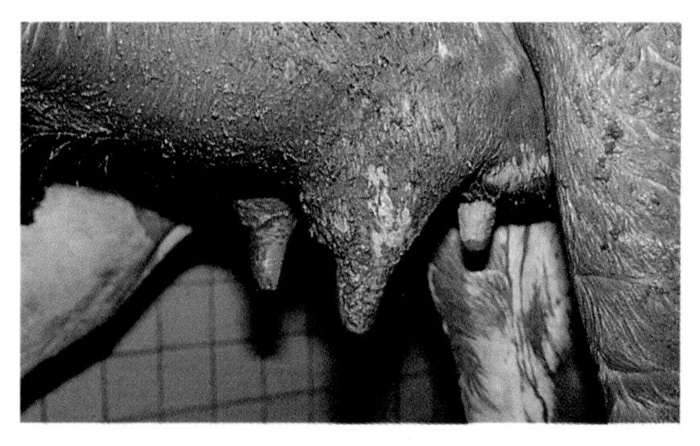

PLATE 6.6 If the teats are dirty they must be washed and dried.

the cows are clean as they enter the parlour is to make sure that they are kept in a clean environment. This is especially important during the housing period.

Hairy udders, as shown in Plate 6.4, are likely to trap dirt and this adds to the amount of work the milker has to do. They should be clipped.

In general, if the cow enters the parlour with visibly clean teats then dry wiping the quarters with a paper towel will suffice (see Plate 6.5). If the teats are dirty, as shown in Plate 6.6, they must be washed **and dried**. Grossly contaminated teats should be soaked before washing to allow the dirt to soften. This allows easier removal of soil. Poor washing procedures assist the spread of bacteria rather than removing them.

Washing of teats is best carried out using water from drop hoses. Warm water is recommended as it is more comfortable for both the cow and the milker. However, if warm water is used, it is essential that header tanks are kept clean and covered, and preferably that a sanitiser is added to the water.

Contaminated water can be a source of *Pseudomonas* infections, (see page 39). Washing teats in the winter with cold water can reduce the milk let-down reflex. It may also have an adverse affect on teat condition. Some people wash the teats and udder with a power hose as the cows enter the parlour. This is not recommended as it must be painful to the cow and soaks the udder as well as the teats.

It is important to ensure that **only** the teats and not the udder are washed otherwise when the milking unit is attached water runs down the wet udder and collects around the top of the liner as shown in Plate 6.7. This is commonly referred to as '**magic water**' because one moment it's there and the next it's gone! If sucked in through the top of the liner at best it contaminates the milk (causing increased TBCs), and at worst causes liner slip, creating impact forces. Impact forces increase the risk of new mastitis infections (see page 63) as there are likely to be high levels of *E. coli* and *Streptococcus uberis* in magic water. The less water used on teats the better. If washed, it is **essential** that teats are dried prior to attachment of the cluster.

The addition of even low levels of disinfectant, for example 60 ppm of iodine or 200 ppm of sodium hypochlorite, is beneficial. It helps to keep the warm water and pipelines free from bacterial contamination. It also reduces the number of bacteria on the teat and helps to keep the milker's gloves or hands clean during the milking process.

In hot climates such as the desert areas of the United States and the Middle East, automatic teat washing in sprinkler pens may be used. Here there are two collecting yards before cows enter the parlour. The first yard is fitted with sprinklers at ground level which jet water up against the udder and teats to remove any dirt, as shown in Plate 6.8. The cows are then allowed to stand while the udders and teats drip dry, before they enter the second collecting pen from which they have immediate access to the parlour. Sprinkler pens reduce the amount of washing necessary in the parlour and so speed up milking. However, as they wet both the udder and teats it is **vital** that the cows are

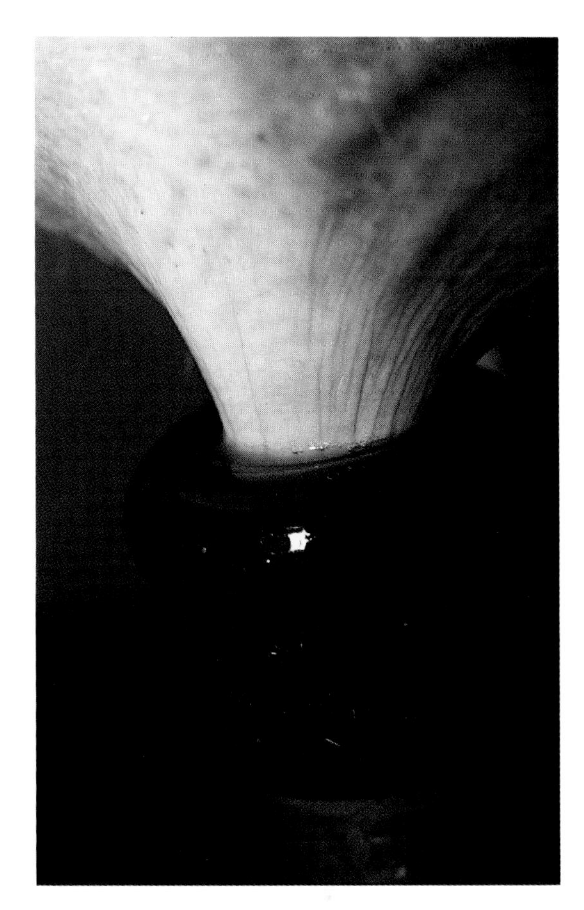

PLATE 6.7 When the udder gets wet, water drains down and collects around the top of the liner and the teat. This is commonly referred to as 'magic water'.

PLATE 6.8 Automatic teat washing sprinklers jet water up against the udder and teats to remove any dirt.

thoroughly dry before entering the milking parlour. The system can only be used in hot climates.

Washing but not drying teats before milking will increase the bacterial contamination in milk and this will raise TBCs. It also deposits bacteria in suspension on the teat end and in so doing increases the risk of environmental mastitis. Finally, excessively wet teats increase liner slip. In herds that wash but do not dry teats before milking the coliform count, which is a measure of the level of environmental contamination of milk, is high (see page 134).

DRYING TEATS BEFORE MILKING

Cows should be dried with a single service paper towel or cloth before the milking units are attached. Under no circumstances should a communal udder cloth, as shown in Plate 6.9, be used as this only spreads mastitis from cow to cow. Udder cloths often become so grossly contaminated that they are virtually impossible to disinfect. The only acceptable udder cloth is an individual cloth per cow that is washed, disinfected and dried between milkings. The cost of this practice is unlikely to be economic in the United Kingdom, although it is practised in some of the large dairies in the United States where energy costs are cheaper.

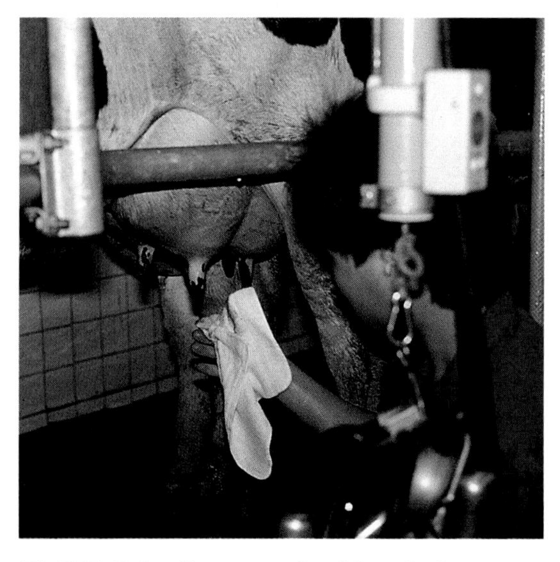

PLATE 6.9 Communal udder cloths spread mastitis bacteria from cow to cow.

The risk of spreading infection through the use of a communal udder cloth cannot be underestimated. Many farmers are convinced that because their udder cloths are placed in a bucket of disinfectant solution during milking all the organisms are killed. Trial work has shown that *Staphylococcus aureus* can survive on udder cloths soaked in disinfectant solutions for 3 minutes. *Streptococcus agalactiae* can survive on cloths for 7 days and can be isolated after soaking for up to 5 hours in a two per cent hypochlorite solution. This is much longer than the few minutes they would be soaked in disinfectant solutions during milking.

It is very simple to show how ineffective disinfectants are in killing organisms on udder cloths. This can be demonstrated to farmers who are convinced that use of cloths poses no risk in spreading infection during milking. Get the milker to squeeze some liquid from an udder cloth onto a blood agar plate during milking. Incubate this for 24 hours in the laboratory. The growth of bacteria shown in Plate 6.10 comes from one such udder cloth. The results speak for themselves.

Paper towels are cheap, disposable and are the ideal choice for drying teats. In some countries old newspapers are used! It is important that a separate piece of paper towel is used on each cow or else you may smear dirt or infection from the first cow onto the second and so on.

Medicated towels are recommended by some people. These are towels impregnated with a disinfectant and are designed for multiple use. They are only intended to **dry** and disinfect clean teats, **not** to clean and dry dirty teats. If they are used to wash dirty teats, there will be little if any benefit over the use of paper towels.

One worry with medicated towels is the possible spread of infection between cows. There may be insufficient concentration of disinfectant and certainly insufficient time to kill all bacteria between cows. This will definitely be the case if the towels are not discarded once they become dirty. In addition, as these towels are expensive, they tend to be overused to economise on cost, further reducing their efficiency.

PRE-DIPPING

Pre-dipping refers to the disinfection of the teat before milking. The aim is to reduce the number of bacteria present on the teat before the milking unit is attached. This will greatly reduce the number of environmental bacteria entering the milk, and in so doing will also reduce the risk of environmental mastitis.

Pre-dip solutions must be fast in action. Only preparations with a proven rapid speed of kill (under 30 seconds) will be effective. To get the maximum benefit, clean teats should be coated in pre-dip solution. A minimum contact time of 20–30 seconds is necessary and then the solution must be thoroughly wiped off the teat before the unit is applied. This ensures that no chemical contamination of milk occurs. Many herds have found a marked improvement in mastitis incidence, TBC and teat condition when they pre-dip. Pre-dips are discussed in detail on page 99.

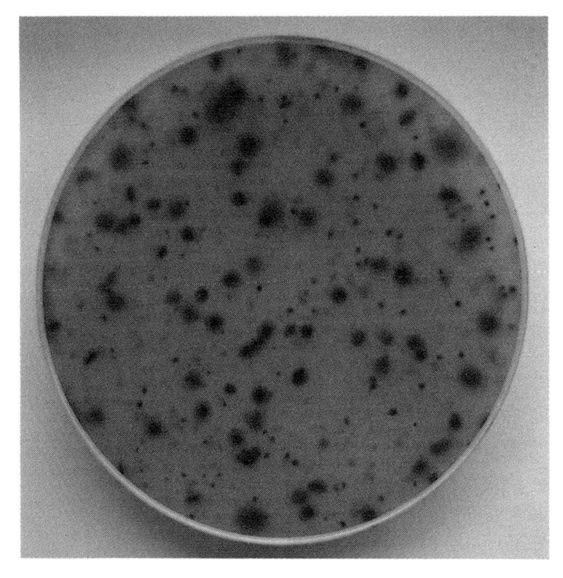

PLATE 6.10 Growth of bacteria from an udder cloth sampled during milking.

MASTITIS DETECTION

Mastitis is inflammation of the udder. The appearance of the milk changes with the type of inflammatory response and it may look clotted, watery or stringy as shown in Plate 6.11. It is not possible to be sure which organism is causing the mastitis simply on the appearance of the milk alone.

The milker may detect mastitis by one or more of the following methods:

- foremilking (see p. 78)
- change in the behaviour of the cow
- observation of quarter swelling
- palpation of the udder
- in-line mastitis detectors
- checking the milk sock or filter at the end of milking

The importance of foremilking has already been discussed. Many herds now rely on the other methods mentioned above.

When a cow comes into the parlour on a different side or at a different time than usual, then this suggests that something is wrong. She may be sick or it could be due to other factors such as bulling. Good stockmanship can help identify early cases of mastitis by picking out cows that act out of character.

There are occasions when the cow has a visibly swollen quarter but the milk appears normal. If the cow is ill, this may be due to toxins produced in the udder by a peracute form of mastitis such as *Staphylococcus aureus* or *E. coli.* This may occur so rapidly that the milk still appears normal. In these cases prompt veterinary treatment will be necessary in order to save

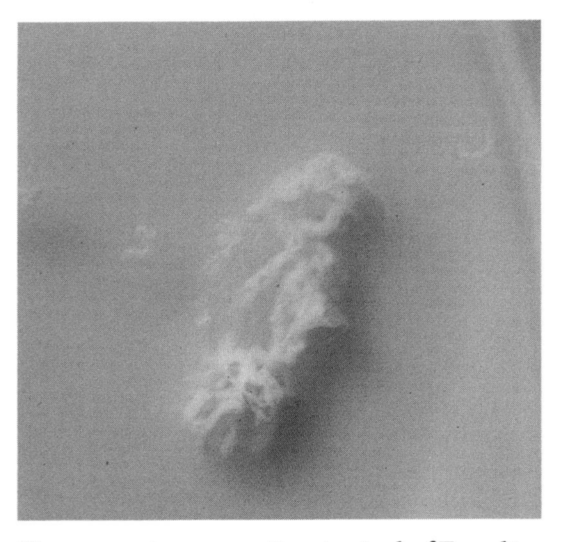

Brown watery secretion typical of E. coli *infections.*

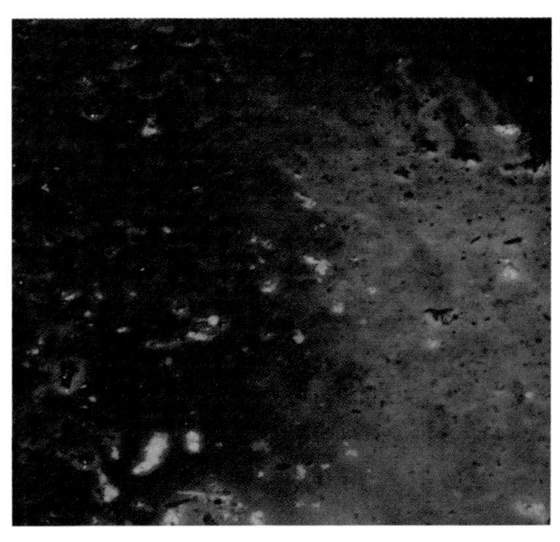

Watery milk with some clots.

Viscous red/brown secretion is associated with gangrenous mastitis.

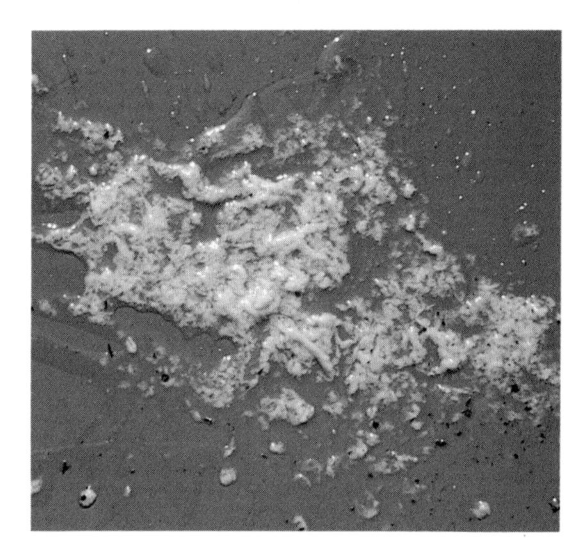

Clotted milk indicating mastitis.

PLATE 6.11 Various types of mastitic milk.

the cow's life. Udder palpation is useful in identifying hot quarters which are often also inflamed and suggest mastitis (see Plate 6.12). On other occasions, there is no swelling in the quarter and the milk shows very little change but the cow is again very sick due to *E.coli* mastitis toxaemia.

In-line mastitis detectors are fitted to the long milk tube (Plate 6.13). They have a wire mesh filter through which most milk passes. Any clots present clog the filter. Milk should be able to bypass the filter without impeding milk flow. Filters should be located in the long milk tube

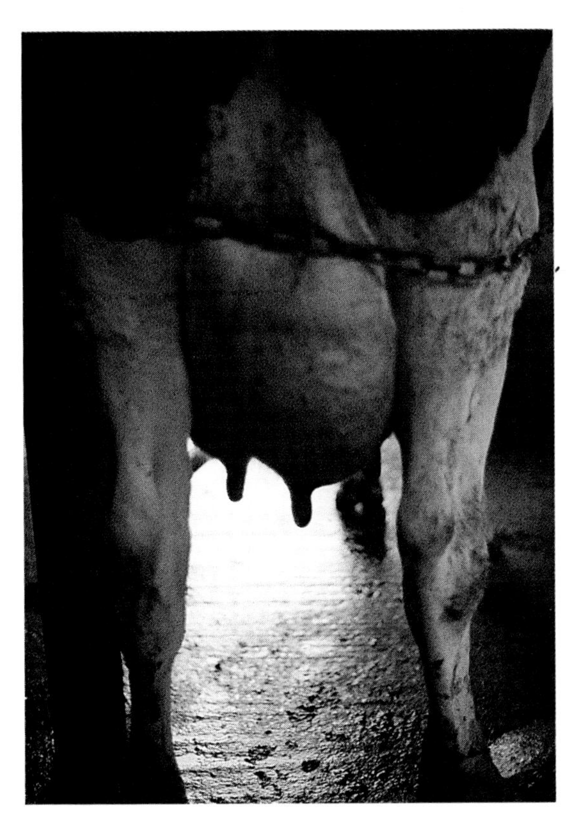

PLATE 6.12 Udder palpation is useful when quarters become hot and inflamed as shown.

PLATE 6.13 In-line mastitis detectors are fitted to the long milk tube and if clots are present they clog the filter.

either at eye level or close to the clawpiece so that they are easily seen as the units are removed. All too often, they are sited in a location where they are difficult if not impossible to observe. These detectors can give a false sense of security to milkers. Some assume that all cases of mastitis can be discovered using this method, even if filters are never checked! In-line filters pick up clotted forms of mastitis, but in cases where milk is watery it can pass through the detector and the mastitis may go undetected.

In order to be effective detectors must be examined after each cow is milked. In parlours fitted with recorder jars, they should be checked **before** milk is transferred into the bulk tank. Problems with TBC and cell count do occur when detectors are not checked because mastitis goes unseen and mastitic milk enters the bulk supply. This is particularly the case with direct to line plants which rely solely on detectors for mastitis identification. Here mastitic milk will have reached the bulk tank by the time clots are seen on the filter. However, despite their limitations, in-line mastitis detectors can still be helpful as an **additional** aid in detecting mastitis.

Examining the milk sock or filter after milking is also important, especially to check on the hygiene of the milkers. (The milk sock or filter is located between the milk pump and the bulk tank). The presence of clots or large amounts of faecal contamination as shown in Plate 6.14 indicates a poor milking routine and poor mastitis detection.

Some farmers rely solely on checking the milk filter for mastitis detection. When clots are found, the procedure is to strip out the entire herd at the next milking to identify the infected cows. In some instances milkers stop stripping when one mastitic cow has been identified. This may leave other mastitis cases undetected. On other occasions no mastitis may be found as the cow may have cleared up the infection herself.

PLATE 6.14 The presence of clots or large amounts of faecal contamination in the milk sock indicates a poor milking routine and/or poor mastitis detection.

In all cases of clinical mastitis, it is advisable to collect pre-treatment milk samples for bacteriological testing. This will allow the identification of the types of mastitis present on the farm so that specific control measures can be implemented. Although it is preferable to process samples as soon as they have been taken, they can be stored in the fridge for up to 3 days before going to the lab. It is essential that sterile samples are collected. The correct procedure is described on page 42.

MILKING THE MASTITIC COW

In an ideal world, as soon as a new case of mastitis is detected the cow should be held back and milked last. Unfortunately this rarely happens as it is time-consuming and there may be no facilities for separating and holding back individual animals. Milking infected cows last eliminates the risk of mastitic or antibiotic milk entering the bulk tank. (Don't forget to remove the milk line from the bulk tank first!) It also reduces the spread of mastitis to the rest of the herd via the milker's hands and contaminated liners, and allows the milker more time to treat cows properly without slowing down the milking process. Remember that liner condition is also significant as worn, cracked liners will retain more bacteria than smooth liners, see page 56.

It is important to disinfect the cluster after milking mastitic cows. This will reduce the risk of transferring infection to other animals. Many milkers dip the units in a disinfectant solution for a few seconds. While this reduces the number of bacteria present, it does not kill all the organisms nor does it fully eliminate the possibility of spread.

Table 6.1 compares the efficiency of different methods of disinfecting clusters. It can be seen that a cold water flush for 5 seconds is ineffective in removing many bacteria. You need to circulate water at 74°C for 3 minutes in order to disinfect the cluster, but this is impractical during milking. The only effective method to sterilise the cluster during milking is to flush it with the water at 85°C for 5 seconds. Most farmers disinfect the cluster by immersing it in a solution of

Table 6.1 Disinfection of teatcup clusters after removal from cows with mastitis. (*22*)

Treatment	*Time*	*No. tested*	*% Clusters positive after treatment*	*No. Staph. aureus/ml recovered per cluster*
Cold water flush	5 s	19	100	100,000–800,000
Circulation of cold hypochlorite (300 ppm)	3 min	19	100	50–2,000
Circulation of water @ 66°C	3 min	18	22	0–80
Circulation of water @ 74°C	3 min	85	0	0
Circulation of water @ 85°C	5 s	530	3	0–15

PLATE 6.15 Back flushing units disinfect the milking unit by passing water at 85°C through the liners.

hypochlorite or iodine solution for several minutes. While this does not sterilise the unit completely, it does remove the majority of the infection, minimising the risk of cross-contamination.

In some of the large dairy herds in the United States where highly contagious *Mycoplasma* mastitis poses a real threat to udder health, back flushing units are fitted to parlours. These disinfect the milking unit by passing water at 85°C through the liners as shown in Plate 6.15. These units are very expensive and provided good mastitis control measures are used, little benefit would be gained from them under UK conditions where *Mycoplasma* infections are rarely encountered.

The simplest and most practical way to reduce the spread of infection is to have a separate cluster for milking mastitic cows. This means that they can be milked as and when they enter the parlour. This cluster can then be adequately disinfected between uses without slowing down the milking process. Many farmers have a separate cluster connected to a dump bucket, as shown in Plate 6.16. By collecting mastitic milk separately, there is no risk of antibiotic contamination.

The dump bucket used on mastitic cows is also often used to collect colostrum from fresh calvers. Freshly

PLATE 6.16 A separate cluster connected to a dump bucket will reduce the risk of spreading infection to clean cows and eliminate the risk of antibiotic residues entering the bulk supply.

calved cows are very susceptible to infection as their resistance to disease is low at this time (see page 24). If the cluster is not disinfected between uses, this may act as a source of new mastitis infections. It is essential that these units are not neglected, that liners are changed frequently and that they are thoroughly cleaned after each milking.

Many farmers still milk mastitic cows into recorder jars and then drain this milk into a bucket or onto the floor. There are several dangers in doing this. The milker may forget to dump the milk so antibiotic residues contaminate the bulk supply. The valve at the bottom of the recorder jar may be faulty and milk may leak past the valve and into the bulk tank. Finally, antibiotics concentrate in butterfat and so jars that are not rinsed out thoroughly after milk is released may still result in antibiotic contamination from butterfat rings around the inside of the jar. Antibiotic residues are discussed on page 157.

MILK LET-DOWN REFLEX

Milk is extracted from the udder by applying vacuum to the end of the teat. This literally 'sucks' the milk out. It is important that the teat does not collapse during milking. This is achieved in two ways: firstly, the venous plexus at the base of the teat (see page 10) becomes engorged with blood and this makes the teat become erect. Secondly, there is an increase in pressure within the udder during milking which causes the teat to become full and turgid, thus making milk removal easier. Release of the hormone oxytocin results in increased udder pressure. Stimulation of the udder and teats causes the pituitary gland at the base of the brain to secrete oxytocin. Oxytocin acts on the alveolar muscles in the udder which then squeeze milk into the ducts. This results in a pressure build-up and produces the syndrome called milk let-down.

Oxytocin release occurs through two forms of milk ejection reflexes: 'conditioned' and 'unconditioned'. Conditioned reflexes are those that the cow takes in through her eyes, ears and nose such as the sound of the vacuum pump and the smell of cake. Unconditioned reflexes occur as a result of teat stimulation such as washing, foremilking and drying.

The level of the conditioned reflex generally remains constant and so the aim of a good milking routine is to maximise the level of the unconditioned reflex. This is where the benefit of a good milking routine pays dividends. Experimental evidence shows that cows milked with a consistent milking routine milk out faster than cows milked with a variable routine, and at the end of milking less residual milk is left in the udder (see page 90).

The key factor in the speed of milkout is not the level of oxytocin in the blood but rather the timing of the oxytocin release. Studies have shown that teat or udder massage for 30–60 seconds immediately before units are attached results in faster milk flow rates. The oxytocin let-down reflex is relatively short-lived and does not extend beyond 10 minutes. Cows should therefore have finished milking within 10 minutes of entering the parlour. Taking full advantage of the milk let-down reflex will result in faster milking, and ensure the most complete removal of milk from the udder, thereby improving production.

UNIT ATTACHMENT

The cluster should be put onto the cow as soon as possible after teat preparation. An efficient milker will attach the units without leaking large amounts of air into the system. This helps to make sure that vacuum levels remain stable and reduces the risk of liner slip and impact forces.

Units should be carefully aligned so that the cluster sits comfortably on the udder without twisting. This ensures that the cow will milk out evenly. Clusters that twist are uncomfortable on the cow, may result in a poor milkout of a quarter and will increase the risk of her kicking off the unit (see Plate 6.17). They will also increase the risk of air being admitted through the top of the liner.

Units are designed to be attached with the long milk hose extending out through the hind legs or forwards towards the head of the cow. In the latter case a support bar, as shown in Plate 6.18, is needed to avoid the unit twisting on the udder. Although commonly used in Europe and the United States, support arms are not popular in the UK.

On occasions, milkers have been known to place stones or bricks on top of

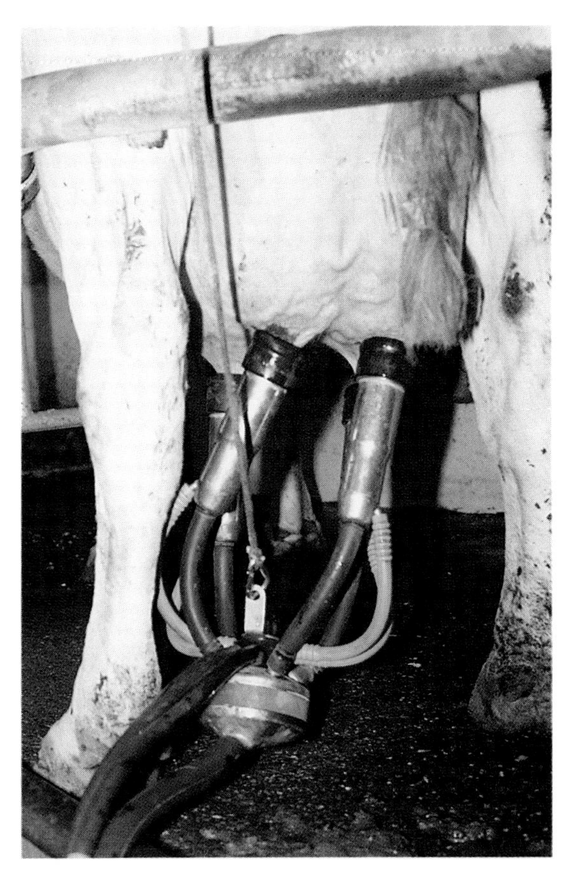

PLATE 6.17 Poor unit alignment. Clusters that twist are uncomfortable on the cow and may result in a poor milkout of a quarter.

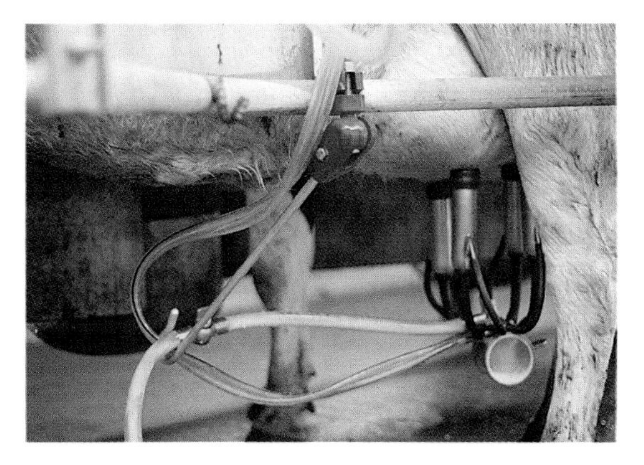

PLATE 6.18 A support arm helps to prevent the unit twisting on the udder by taking the weight of the long milk hose.

the clawpiece to try to speed up a slow milker, as shown in Plate 6.19. This practice is not to be encouraged as it causes excessive pulling on the teats which increases any teat damage already present and may indirectly slow milking down even further.

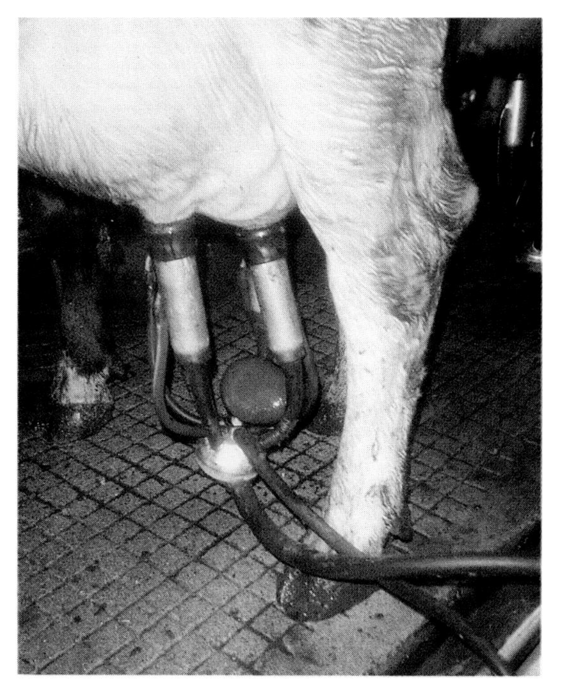

PLATE 6.19 Milkers have been known to place stones or bricks on top of the clawpiece to try and speed up a slow milker. This is not recommended.

DISINFECTION OF CLUSTERS BETWEEN COWS

At one stage dipping clusters into a bucket of disinfectant solution between cows was a popular procedure in the milking routine. It was generally believed that dipping units in a solution of hypochlorite for a few seconds would remove all bacteria from contaminated liners. The difficulty of disinfecting clusters has been shown in Table 6.1.

Although cluster dipping will help to reduce bacteria numbers, it is unlikely to be largely effective in eliminating the spread of bacteria from cow to cow. The time spent on this procedure could be put to more beneficial use in other areas relating to the milking routine.

RESIDUAL MILK

No matter how long you leave the milking unit on the cow, not all milk will be removed from the udder. The remaining milk is called residual milk. The amount of residual milk is usually 0.5 l for heifers and 0.75 l in cows, but may be as high as 15 to 25% of the total amount of milk present in the udder before milking.

There are a variety of factors that can increase the amount of residual milk:

- disturbing or frightening cows just before or during milking which will affect the let-down reflex.
- delay between udder stimulation and attaching the teat cups.
- irregular milking intervals.
- teat injuries.
- poor unit alignment leading to incomplete milkout in one or more quarters.
- poor ACR adjustment leading to early cluster removal.

A good milking routine will help to keep the amount of residual milk to a minimum. If a large amount of milk remains in the udder this may aggravate subclinical mastitis, especially with *Streptococcus agalactiae* infections. It will also decrease milk yield by 5 to 10%.

Machine Stripping

Machine stripping is the process whereby downward pressure is applied to the clawpiece with one hand, while the quarters are massaged with the other as shown in Plate 6.20. The intention is to maximise milk removal and reduce the amount of residual milk in the udder.

There is a danger that if machine stripping is carried out with great vigour and enthusiasm, air will be sucked in through the top of the liner resulting in impact forces. Impact forces may cause a massive reverse flow of milk against the teat end (see page 63). If bacteria penetrate the teat canal a new infection may become established. For these reasons machine stripping is not recommended. Today's dairy cow is selected for good udder and teat conformation. In addition, the ability of the milking machine to reduce the amount of residual milk has greatly improved with modifications to its design.

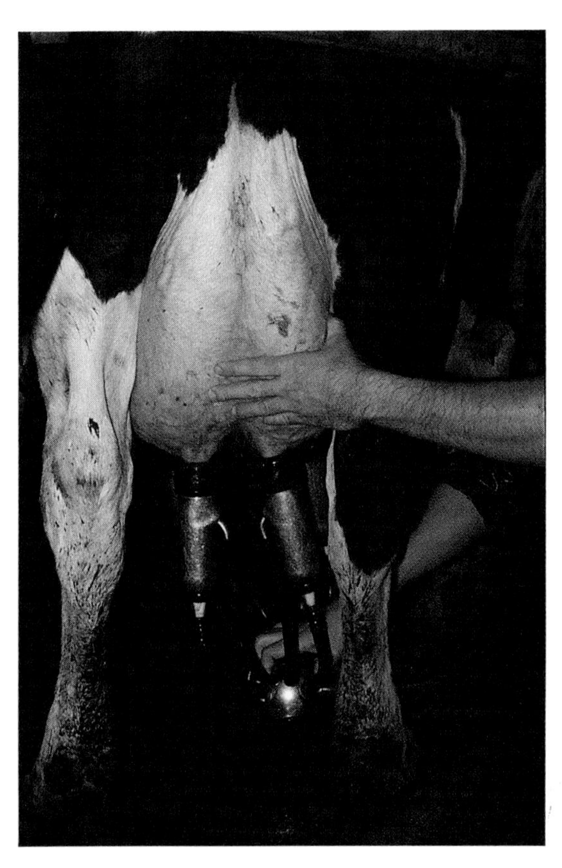

PLATE 6.20 Machine stripping is the process whereby downward pressure is applied to the clawpiece with one hand, while the quarters are massaged with the other. This is not recommended.

UNIT REMOVAL

Once cows are milked out, the vacuum supply to the cluster should be shut off. Atmospheric air enters the clawpiece through the air bleed hole releasing the vacuum and allowing the unit to 'fall off' the udder. Where ACRs (automatic cluster removers) are fitted, they should operate in the same way. A cluster removed while still under vacuum can result in large impact forces and may cause teat sphincter damage. (See Plates 14.25 and 14.26.)

Overmilking is not to be encouraged. While it does little harm in itself, providing of course that the machine is functioning correctly, it increases the unit time on and so slows down milking. As liner slip occurs most frequently towards the end of milking, overmilking in a poorly functioning parlour may well increase the number of new infections.

POST-MILKING TEAT DISINFECTION

Immediately after the cluster is removed the entire surface of each teat should be coated in a disinfectant solution as shown in Plate 6.21. This can be applied as a dip or as a spray.

The aim of post-milking teat disinfection is to kill any bacteria

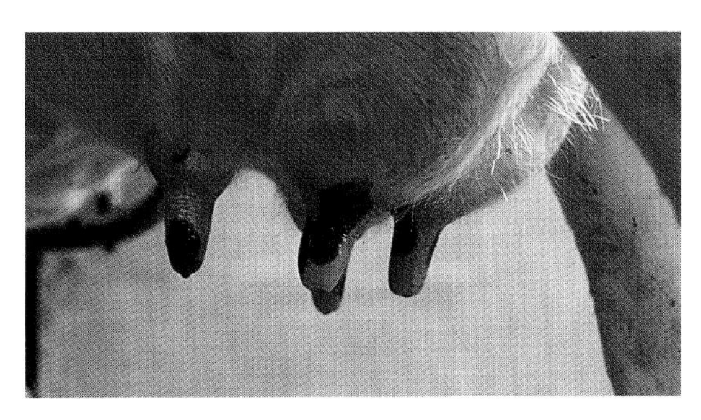

PLATE 6.21 Immediately after the cluster is removed following each milking, the entire surface of each teat should be coated in a disinfectant solution. Note the yellow stain left by the iodine in the dip.

transferred onto the teat during milking, before they have a chance to colonise or infiltrate the teat canal. Post-milking teat dipping is an **essential** method of controlling contagious mastitis. It is less effective against coliform and other environmental forms of mastitis (for which pre-dipping is more important). Teat dips are discussed in Chapter 7.

After cows leave the parlour they should have access to food and water so that they will remain standing for between 20 and 30 minutes while the teat canal closes fully. If cows lie down immediately after milking while the teat canal is open, environmental bacteria can easily penetrate the udder and mastitis may result (see page 17).

MILKING ORDER

The order of milking is very important if a herd is divided into groups.

To help reduce the spread of mastitis cows should always be milked in the following order:

- fresh calvers (most susceptible to mastitis)
- high yielders
- medium yielders
- low yielders
- high cell count cows
- mastitic and other treated cows

Fresh calvers should be free from infection as dry cow therapy will have eliminated most of it in the dry period. Late lactation cows are likely to have an increased amount of subclinical infection due to longer exposure to mastitis organisms throughout lactation. Therefore, the greatest risk of contagious mastitis will come from late lactation cows.

FREQUENCY OF MILKING

There is a higher yield when the milking frequency is increased to three or four times a day, without significant effect on milk composition: yields can be increased by up to 20% in heifers and 15% in cows. This is due to the removal of the milk inhibitor protein, resulting in more milk being synthesised in the udder (see page 15). In addition, as milk is removed from the udder more frequently, flushing out any bacteria in the teat canal and udder (and despite a possible increased risk of new infections from extra milking) the mastitis incidence in herds that milk three times a day tends to be lower than twice daily counterparts.

SUMMARY OF THE MILKING PROCEDURE

The aim of the milking routine is to milk clean, dry teats with a correctly functioning milking machine as efficiently as possible, thus posing minimal risk to udder health while maintaining milk quality. This is achieved by the following routine:

- foremilk cows
- dry wipe clean teats but wash and always dry dirty teats
- pre-dip: allow a 20–30 second contact time and wipe off
- attach the milking unit within one minute of teat preparation
- check machine alignment
- when the cow is milked out, shut off vacuum and then remove the cluster
- coat teats with teat dip as soon as the cluster is removed
- allow cow to exit to a sheltered yard with access to food and water so that she can remain standing for 20–30 minutes

The additional time spent diligently carrying out these procedures will be rewarded by a lower incidence of mastitis, cleaner milk production, increased milk yield, increased satisfaction for the milker and greater comfort for the cow.

CHAPTER SEVEN

Teat Disinfection

Teat disinfection is one of the most important preventive measures in mastitis control. This chapter examines the reasons for teat disinfection, the methods of application (dip or spray), the chemicals used, some of the associated management faults and the importance of chemical residues.

Teat disinfection can be carried out immediately prior to milking or straight after milking.

PRE-DIPPING

The disinfectant is applied just before milking, and teats must be wiped clean of disinfectant before the cluster is attached. Pre-dipping is a relatively new concept, aimed at reducing the incidence of environmental mastitis and reducing TBC. As environmental mastitis is mainly a winter housing problem, some herds only pre-dip during the winter.

POST-DIPPING

The disinfectant is applied as soon as the milking unit is removed. Teats do not need to be wiped dry after the post-dip has been applied. Post-dipping is of major importance in the control of contagious mastitis (see page 29) and should be carried out in every herd, at every milking, throughout the year.

METHOD OF APPLICATION: DIP OR SPRAY

Because of the importance of removing bacteria from all the teat (see page 17), it is essential that the **whole teat** and not just the teat end, is disinfected. This is likely to be best achieved by dipping, although spraying can be effective if carried out conscientiously.

Dipping

Dipping uses less product than spraying (approximately 10 ml per cow per milking

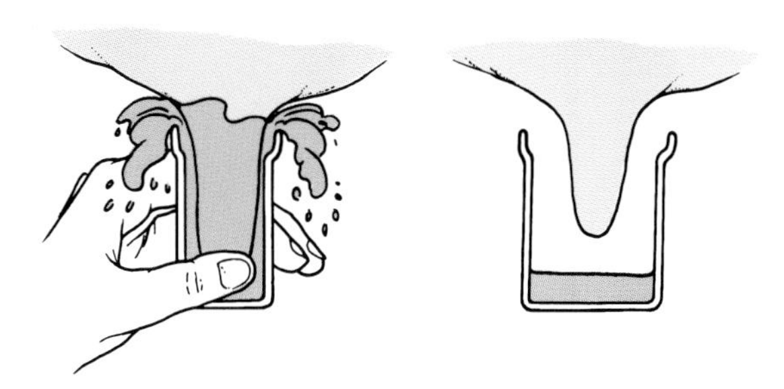

FIGURE 7.1 (Left) A small, narrow teat dip cup which is overfull will cause the iodine to be wasted and the milker's hands to become stained, whereas an excessively wide cup with only a small quantity of teat dip (right) may mean that small teats are not adequately dipped.

for dipping versus 15 ml for spraying) and, provided that it is carried out correctly, can provide excellent teat cover. The pot should be large enough to contain the teat without excessive spillage of dip and, at the same time, it should be full enough to ensure that small teats reach the dip solution (see Figure 7.1).

Dual compartment anti-spill cups are also available (Figure 7.2). When the bottom compartment is squeezed, dip is forced into the top. If the pot should then be knocked over (or kicked out of the milker's hand) it is only the dip in the top of the pot which is spilled. These cups often have a hook on the side, allowing them to be attached to the milker's belt (Plate 7.1) and therefore readily available for use. Whatever type of pot is chosen, it should be applied so that the rim makes contact with the udder and then shaken to ensure total teat cover.

Teat dip pots should be cleaned regularly, to prevent contamination. Any dip remaining in the pot at the end of milking should be discarded and the pot cleaned before re-use at the next milking.

PLATE 7.1 Dip: pots of pre-dip and post-dip conveniently attached to the belt during milking.

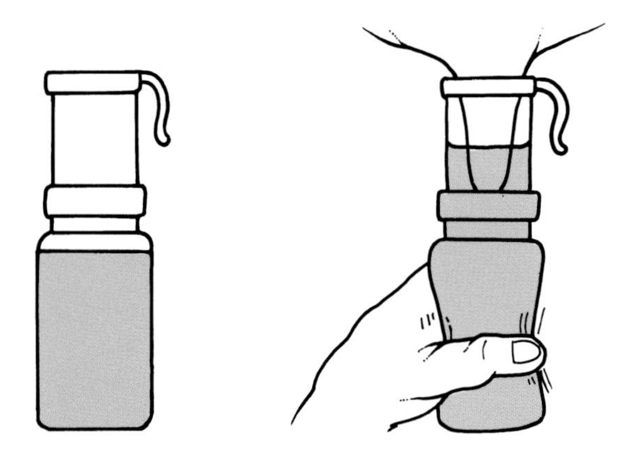

FIGURE 7.2 Anti-spill cups are available. Only the dip in the top chamber is spilled if the cup is knocked over.

Spraying

Teat spraying can also be effective, but must be carried out conscientiously. It is much easier to achieve only partial cover than with dipping. Spray lances should be of reasonable length, with the nozzle pointing upwards and not directly out from the end (Figure 7.3). Spray should be applied from beneath the teats whilst rotating the lance in a circular action below the base of the udder. At least two rotations will be needed to achieve full cover.

In herringbone parlours many milkers open the gate, releasing the cows, and then try to apply teat spray as they walk past. Unfortunately this results in only partial teat cover. If iodine disinfectants

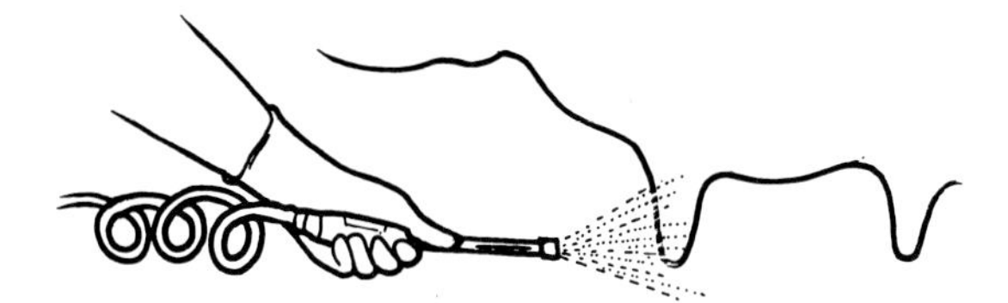

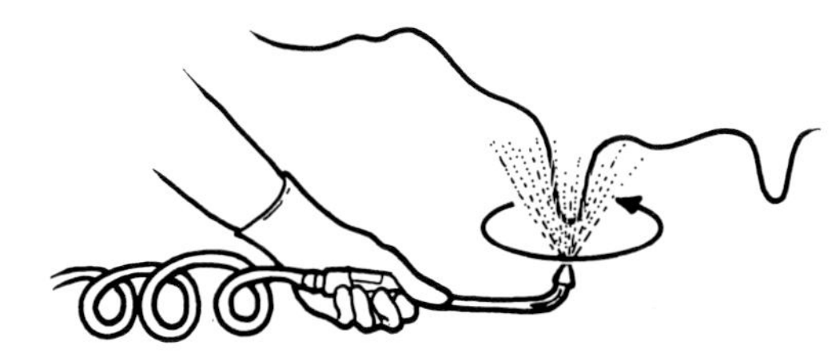

FIGURE 7.3 Spray applied from the side only achieves partial cover. Spray should be applied from the bottom of the teat in a circular motion to ensure total coverage.

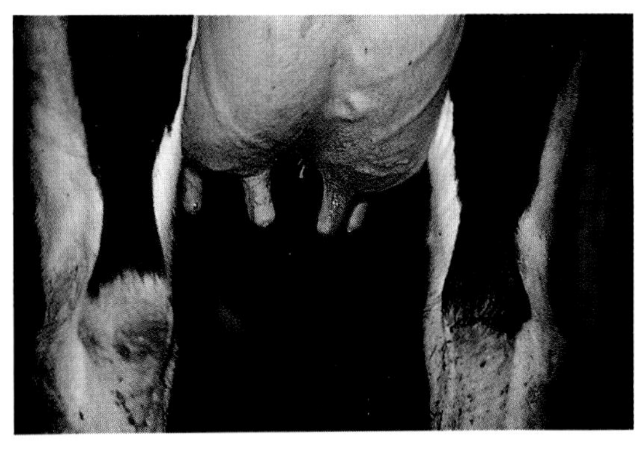

PLATE 7.2 Poorly applied spray leads to only one side of the teats being covered.

are used it is easy to see when only one side of the teat has been coated, as in Plate 7.2. The disinfectant may well run to the end of the teat, thus eliminating teat-end colonies and reducing the most important aspect of mastitis transmission. However, the absence of disinfectant on one side of the teat could allow the establishment of a reservoir of mastitis pathogens on the untreated teat skin.

Spray nozzles should be regularly checked as they can become partially blocked and produce poor teat cover.

A comparison between teat dipping and spraying is shown in Table 7.1.

Table 7.1 Comparison between dipping and spraying

	Dipping	*Spraying*
Teat cover	generally good	good if careful
Volume used per cow/milking	10 ml	15 ml
Cost	very cheap equipment	more expensive to install
Points to watch	dirty teat dip cups	blocked nozzles causing slow flow rates
	keep pot full	
	cows with very short or long teats	solution running out during milking

PREPARATION AND STORAGE OF DIPS

Some dips are bought ready to use, while others are supplied as concentrates and have to be diluted. Ready-to-use products are often more stable, as they have been carefully formulated. When diluting a concentrate the instructions should be followed closely and, ideally, only enough solution for a few days should be made up to avoid deterioration.

Unused containers of dip should be stored away from cold areas, since freezing may lead to separation of water from the chemical. When in use, make sure that the top of the drum is not left open in an area where large quantities of water are splashed. Water contamination will dilute the teat dip or, even worse, if contaminated by circulation cleaner rinse water, the dip may become denatured and ineffective.

CHEMICALS USED IN POST- AND PRE-DIPPING DISINFECTANTS

There is a range of chemicals used for both teat dips and sprays. The most common are the following:

Iodophors

These are probably the most widely used compounds in dips. They consist of iodine in association with a complexing agent which essentially acts as a 'reservoir' of inert iodine. As the free 'active' iodine is slowly used up by reacting with (and killing) bacteria, more free iodine is released from the complexing agent reservoir. This process is able to maintain a constant level of active ingredient in post-dips of around 0.5–1.0%. Iodophors are formulated in an acid solution which can be irritating to teat skin, hence most products incorporating them contain significant levels of emollients (see page 98). Like other teat dips, iodophors are not selective in their action. They will react with any organic matter and so if teats are badly soiled or heavily coated with milk, or if the teat dip cup becomes contaminated with faecal material then the efficacy is markedly reduced.

One advantage of iodine is its colour. It stains skin and so it is easy to see how well teats have been covered after milking (though the stain left on the herdsman's hands may not be appreciated!). Iodine dip which has been excessively diluted looks 'pale' in the pot but can still cause staining. Some milkers dislike the smell of iodine and inhalation of the fumes produced may cause unacceptable respiratory irritation, especially when teat spraying.

Quaternary Ammonium Compounds (QUATs)

These teat dips consist of the quaternary ammonium compound (the bacteria killing component), a 'wetting agent' to assist in greater penetration of skin and dirt, pH buffers to stabilise the acidity of the product, emollients and water. Colouring agents may be added to show that the teats have been dipped and thickening agents may give increased persistence on teat skin. Quaternary ammonium compounds are not irritant to teat skin, although careful formulation is necessary to maintain efficacy. Effectiveness against *Pseudomonas* and *Nocardia* is very doubtful and these bacteria have even been known to grow in QUAT solutions!

Chlorhexidine

Commonly used at 0.5%, chlorhexidine has a wide activity against most bacteria and is less affected by organic material than the other disinfectants. Emollients are needed to protect against skin irritation.

Hypochlorite

Hypochlorite is by far the cheapest product available. Its main disadvantage is that it rapidly reacts with organic material (milk, faeces and skin debris) and becomes ineffective. Used at the usual concentration of 4.0% it can also irritate the milker's hands, cause damage and bleaching of clothing, and may result in quite marked drying of teats especially when first used. These effects are partly caused by the inclusion of sodium hydroxide (around 0.05%), which is sometimes used to stabilise the product.

Ideally, hypochlorite should be introduced at a low concentration and then slowly built up to 4.0% (40,000 ppm). Provided that weather conditions are favourable, teat skin often adapts well and the product can be used without severe reaction. There are anecdotal reports that its strong oxidising action improves the rate of healing of teat-end damage (e.g. black spot, page 177) and of viral skin lesions such as pseudocowpox (page 173).

Due to formulation problems, if emollients are used they must be added immediately prior to milking. Hypochlorite solutions are relatively unstable. They should be stored under cool conditions and with the lid closed, otherwise they can evaporate quite quickly and lose their potency.

There are hypochlorite derivatives available, for example 5 g per litre sodium dichloroisocyanurate, which are more stable and have a less severe skin-drying effect.

Dodecyl Benzene Sulphonic Acid (DDBSA)

Used at 2.0% inclusion, DDBSA dips are nonirritant to teats and to the operator. They have a wide range of activity against most bacteria but are ineffective against bacterial spores. They have a longer length of action than some dips (and hence may confer some protection against coliforms) and work quite well in the presence of organic matter.

The best dip or spray is the one seen dripping off the cow's teats as she leaves the parlour!

As post-dipping is more common than pre-dipping, it will be discussed first.

POST-MILKING TEAT DISINFECTION

There are three major reasons for carrying out post-milking teat disinfection, namely:

- removal of mastitis bacteria from the teat skin.
- removal of bacteria from teat sores.
- improving teat skin quality.

Removal of Mastitis Bacteria

When a cow with an infected quarter is milked by machine, milk from the infected quarter can pass through the claw, reflux up through the short milk tube and infect other teats (page 63). In addition, some mastitis bacteria remain on the liners and this infection can be transmitted to the next 6–8 cows to be milked (see page 32).

Although mastitis bacteria sometimes penetrate the teat canal by rapid reflux of milk (teat-end impacts, see page 63) they are normally deposited at the teat end, on the outside of the teat canal. However, unless removed, the bacteria start to multiply (to form colonies) and then slowly grow up through the teat canal. It is the adhesive properties of the contagious mastitis organisms (see page 29) which allows them to do this. Once they have passed through the teat canal and into the udder, a new infection may be established.

Post-milking teat disinfection removes bacteria deposited during the milking process and as such it is an extremely important control measure against contagious mastitis. The disinfectant should be applied **as soon as** the cluster is removed. At this stage the canal is still open (see page 17) and so some disinfectant can penetrate the teat orifice. This ensures that those mastitis bacteria which have started to enter the canal will also be killed.

Removal of Bacteria from Teat Sores

Any skin lesion which is infected will be slow to heal. Teat disinfection removes bacteria from the skin surface. This promotes healing and maintains teat skin in optimum condition. Rough, cracked or chapped teat skin can be a reservoir for organisms such as *Staphylococcus aureus* or *Streptococcus dysgalactiae*. Thorough disinfection of the whole teat is important to ensure that **all** bacteria are killed.

Improving Skin Quality with Dip Additives

Teat skin has relatively few sebaceous glands (see page 12), and so continual washing followed by exposure of damp teats to a cold and windy environment can remove protective fatty acids and lead to cracking. The most common additives to teat dips are:

- emollients: they form a seal around the

skin to prevent further water loss by evaporation. Similar products are used in udder creams.
- humectants: these assist in drawing water into the skin, examples are lanolin and glycerine.

Lanolin and glycerine are the most common additives and may be included at up to 10% concentration of the dip. As the level of additive increases, the proportion of disinfectant and hence the bacterial killing ability of the final product decreases (see Figure 7.4). For this reason additives are rarely included above the 10% level. If more additive than this is used, then the product may also become too thick and difficult to put through a spray line.

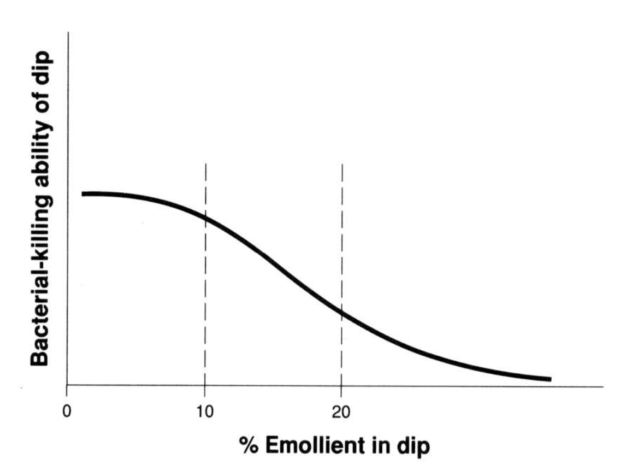

FIGURE 7.4 Emollient levels above 10% significantly depress the bacteria killing ability of the dip.

Automatic Teat Disinfection Systems

Automatic teat disinfection systems, sited at the exit to the milking parlour, are available. The majority are activated by an electronic 'eye'. As the cow walks past, the 'eye' triggers a burst of disinfectant spray from a nozzle or a raised bar on the floor (Plate 7.3) and directs it onto the udder.

Automatic systems are being continually improved, but as yet none are as effective as thorough teat dipping. Their main disadvantages are:

- some cows rush past and may only receive a small amount of spray, or may be missed altogether.
- cows with very high udders may be missed.
- faeces deposited on the spray jetter by one cow could be sprayed onto the following cow.
- some spray systems have a jetter bar which can make contact with (and contaminate) the teats of cows with very pendulous udders.
- if sited outside the parlour, the disinfectant spray may be deflected away from the teats during windy weather.

PLATE 7.3 Automatic teat disinfection system sited at the exit of the parlour.

- as most systems are sited at the parlour exit, they apply the teat disinfectant when the cows are leaving the parlour. This could be some time after the unit has been removed when the teat canal has already started to close. Because of this, high cell counts associated with organisms such as *C. bovis* have occasionally been reported with automatic teat disinfection.
- the system could run out of teat disinfectant without the operator knowing. Alarm devices should be fitted.

These points need to be considered carefully before an automatic teat disinfection system is installed.

Limitations of Post-milking Teat Disinfection

Although widely used and a vital part of every mastitis control programme, post-milking teat disinfection has some limitations:

- it has no effect on existing infections – if teat dipping is introduced into a herd already heavily infected with contagious organisms, you cannot expect a rapid reduction in cell count and mastitis incidence. Although dipping prevents the transfer of bacteria and hence reduces the rate of **new** infections, it has no effect on existing infections. For example, in one trial over a 12-month period, a 50% reduction in new infections produced only a 14% reduction in the overall number of quarters infected. Culling and dry cow therapy are therefore vital additional control measures.
- its main effect is against contagious organisms – environmental infections are thought to be transferred onto the teat end **between** milkings and propelled through the teat canal by teat-end impacts during the milking process. As post-milking disinfectants have a relatively short period of action (e.g. 1–2 hours) after they have been applied, they will have a limited effect against environmental mastitis. Pre-milking disinfection is therefore more important in the control of environmental mastitis.
- it may cause teat irritation – this is particularly the case during wet and cold weather. Some chemicals are more irritant than others, although their adverse effects can be reduced or avoided by the inclusion of emollients. Under subzero conditions, some farmers discontinue teat disinfection. Disinfectants are temperature-sensitive, hence during very cold weather teat dips are not only more irritant, but they also have a lower bacterial killing power.
- it is inactivated by organic matter – all disinfectants are less active in the presence of milk or faeces. For this reason it is important to ensure that any remaining teat dip is discarded from the cup at the end of each milking, that the cup is cleaned and that new dip is added before the next milking.

Seasonal Use of Dips

Some herds have tried stopping post-milking teat disinfection during the summer, only to find that cell counts begin to rise due to a build-up of contagious organisms. In order to be effective, **all** teats must be disinfected after **every** milking.

PRE-MILKING TEAT DISINFECTION

Pre-milking teat disinfection is an important control measure against environmental mastitis. Teats should be clean prior to the application of the milking machine. They may be washed and the use of a sanitiser in the water further reduces bacterial burdens. If teats are washed, then they **must** be dried before milking (page 80).

Although dry wiping or washing and drying helps to reduce bacterial levels on teats, it is by no means as effective as applying a pre-milking disinfectant. Table 7.2 shows the benefits of teat disinfection over and above washing and drying, in an experiment where teats were first exposed to an experimental challenge of *Streptococcus uberis* solution 1–2 hours before milking.

Table 7.2 Comparison of the effectiveness of different methods of pre-milking teat preparations. (*23*)

Teat preparation	*No. of quarters infected*	*% reduction*	*% further reduction*
No preparation	27	–	–
Wash and dry	15	43	–
Wash, dry, then pre-dip and dry	9	67	40

Washing and drying produced a 43% reduction in the percentage of quarters infected. There was an **additional** 40% decrease in infected quarters when teats were dipped in a disinfectant prior to milking, even though previously the teats had been washed and wiped.

Pre-dipping, first introduced in California, is now widely used in North America and is becoming increasingly popular in Europe. Its prime effect is against environmental mastitis. One large field trial in the United States (see Table 7.3), involved four herds over a three-year period. The cows were individually designated 50:50 as pre-dip and control animals, although all were housed, fed and milked as a single group in each herd. Results showed a 46% reduction in the incidence of environmental infections caused by *Streptococcus uberis* and *E. coli* in those cows on which pre-dip was used.

Table 7.3 The effect of pre-dipping on the reduction of new intramammary environmental infections in four commercial dairy herds (*24*)

	Number of quarters at risk	*Number of infected quarters*			*% reduction*
		S. uberis	*coliforms*	*total*	
Control	553	31	41	72	–
Pre-dip	619	18	21	39	46

Results from smaller field trials in the United Kingdom have been less conclusive, with one trial showing a similar 50% reduction in clinical mastitis incidence but another showing a more limited effect.

Pre-dip should be applied immediately before the application of the milking units, and after the teats have been foremilked, washed and wiped. Pre-dip needs a minimum contact time of around 30 seconds and then must be wiped off prior to the application of the machine. Clearly speed of action is important. One novel product with a high (2–3 ppm) free iodine but low (0.1%) total iodine content is currently sold as having a very rapid action, with a 99.99% kill of teat surface bacteria within 30 seconds. This product is also stabilised at a high pH and hence can be used without an emollient.

Not only can pre-dipping reduce environmental mastitis, but if teat contamination is the cause of high TBCs, then pre-dipping can improve bulk milk TBCs. Improving housing conditions to avoid excessive soiling of the teats is also important.

It is certainly logical that if the teat is left soaking in disinfectant for a period of time and then wiped, this must be a very effective method of removing dirt and debris.

Teats which are grossly soiled should, of course, be cleaned first and then pre-dipped. The effectiveness of pre-dipping is seen in some herds where coliform counts in milk may reach zero (see page 45).

As with post-dipping, care should be taken to ensure that the pot does not become contaminated with faeces.

A further claim by the proponents of pre-milking teat dipping is that the teats are more moist and supple when the milking machine is applied and this leads to less liner slip. Some say that teat condition will also improve, but this will clearly depend on the type of dip used (e.g. a high or low emollient). A marked improvement in teat skin condition could explain the anecdotal reports of pre-dipping leading to an improvement in cell count.

A few farmers have used standard post-dip products as a pre-dip, sometimes by diluting the post-dip 50:50 with water. This should be avoided for three reasons:

- post-dips may not have the very rapid speed of kill required of a pre-dip.
- the high iodine concentrations used in post-dips could lead to residues if the dip is used as a pre-dip.
- if a diluted post-dip is used as a pre-dip, it is essential that full strength solution is still retained for post-dipping, otherwise post-dip efficacy will be reduced.

The ideal situation would be to have a single product which could be used as both a pre- and post-dip.

In summary, a comparison of the major points of pre-dipping and post-dipping is given in the following table:

Table 7.4 Comparison of pre-dipping and post-dipping

	Pre-dip	*Post-dip*
Season	particularly during housing periods depending on climate	essential throughout the year
Speed of action	must be rapid	not important
Main effect against	environmental mastitis	contagious mastitis
Effect on:		
Cell count	limited	decreases SCC
TBC	decreases TBC (if dirty teats are contributory)	no effect on TBC

IODINE RESIDUES

Concern has been expressed about the widespread use of iodine products leading to increased milk iodine levels. Milk is certainly an important source of iodine for man. Most milks contain around 350 ug/litre of iodine. As the adult human daily dietary iodine requirement is 150 ug, this would be obtained from the consumption of 430 ml (0.75 pints) of average milk. The majority of iodine in milk (around 70–80%) comes from the cow's diet (see Figure 7.5). Widely differing diets can lead to great variation in milk iodine content. For example, in one trial, herd bulk milk iodine ranged from 200 to 4000 ug/litre. Very small amounts of iodine may come from bulk tank cleaners

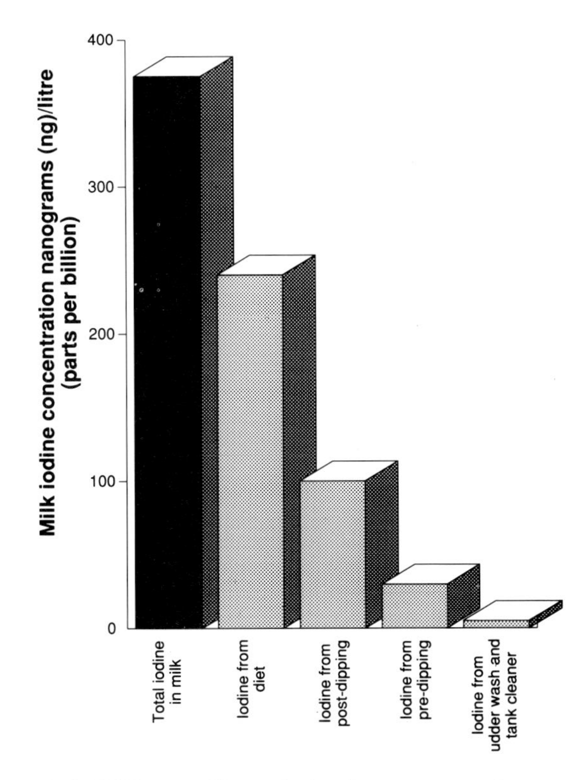

FIGURE 7.5 The relative importance of different sources of iodine in milk. There is an enormous variation between farms in the amount of iodine from the diet.

and possibly sanitisers added to teat washing water. Perhaps surprisingly, more iodine residues are derived from post-milking teat disinfection than from pre-dipping. This is due to a combination of factors:

- pre-milking teat disinfectants are wiped from the teats before the cluster is attached.
- iodine applied immediately post-milking will penetrate the teat canal.
- in herds where teats are only dry-wiped before milking post-dipping iodine residues will still be present on the teats at the next milking.
- there is good evidence that iodine can penetrate skin and then the teat wall and can pass into milk in the teat cistern.
- one novel formulation of pre-dip has a particularly low total iodine content (0.1%) whilst maintaining a high free iodine concentration.

When iodine teat disinfectant is applied by spray, there is an increase in atmospheric iodine. Levels do not reach values high enough to represent a human health hazard, although the vapour may irritate some herdsmen.

Suggested maximum dietary iodine intake limits for the United Kingdom are 2000 ug/day (thirteen times the dietary requirement). A daily intake of only 500 ml/day (less than a pint) of some extreme farm milks would be needed to reach these limits. Most of the milk consumed is from mixed sources, however, and it is therefore unlikely that these extremes would occur with purchased milk.

CHAPTER EIGHT

The Environment and Mastitis

Maintaining a clean and comfortable environment for cows is of major importance in both mastitis control and in the production of clean, quality milk.

ENVIRONMENTAL VARIATION

Dairy cows are kept under a very wide range of conditions. They may be grazing pastures during a dry summer, or plodding through muddy gateways during a wet spring or autumn. They may be housed in open yards (Plate 8.1) in the hotter climates of Arizona, California, Israel or Saudi Arabia, or in cubicles (free-stalls), cowsheds or straw yards (Plate 8.2) in Europe and the more northern parts of North America.

PLATE 8.1 Cows in an open sand yard, typical of hotter climates.

Whatever the environment, the two major factors which can lead to an increase in mastitis and bacterial contamination of milk are:

- housing – confinement leads to much closer cow-to-cow contact and therefore a greater opportunity for faecal contamination.
- humidity – damp conditions facilitate the movement of faeces onto udders and allow greater multiplication of environmental organisms.

Large numbers of *E. coli*, 1,000 (10^3) per gram, are normally excreted in the faeces. This can increase considerably (up to 10^6

PLATE 8.2 Straw yard, typical winter housing for temperate climates. Although very comfortable for the cows, a high degree of management is required to avoid mastitis.

per gram) in a freshly calved cow fed on a high concentrate ration. Even worse than this is the warm mixture of milk, bedding and faeces which can sometimes accumulate at the rear of the cubicle of a high-yielding cow leaking milk (Plate 8.3). Bedding like this may contain up to 1,000 million (10^9) *E. coli* per gram and represents a serious challenge to the mammary gland especially during milk leakage when the teat canal is open and highly susceptible to mastitis.

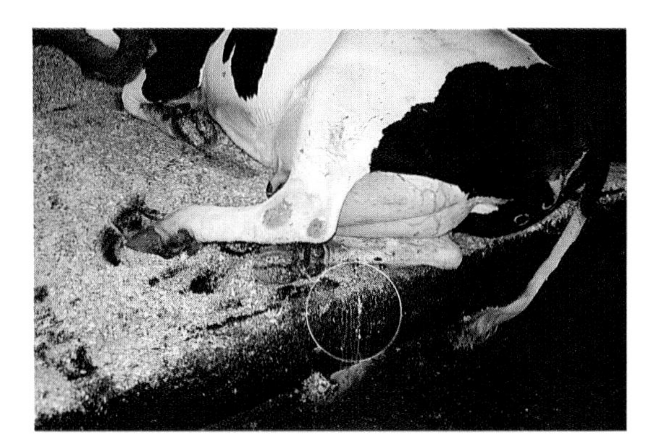

PLATE 8.3 A cow 'leaking' milk onto the cubicle bed. As milk is an excellent bacterial nutrient, the combination of milk, faeces and bedding produces a very severe mastitis risk.

BEDDING TYPE

Both the type of bedding and the way in which the bedding is managed can have a marked effect on coliform levels. This is shown in Table 8.1, which compares four housing systems. It is interesting to note that no cases of coliform mastitis were seen in the 150 cows housed over the winter in systems 1, 2 and 3, (sand, straw and well-stored sawdust) but seven cases occurred in three months in only 24 cows housed in system 4 (damp sawdust).

Different types of bedding seem to support the growth of different organisms and this leads to increased levels of teat contamination by those organisms. This

Table 8.1 Coliform levels using different housing systems (*25*)

Group	*Housing system*	*Number of coliforms/g bedding*	*Cases of coliform mastitis*
1	Sand cubicles	37,000	0
2	Straw yards	47,000	0
3	Well-managed sawdust yards	44,000	0
4	Poorly managed sawdust yards	66,000–69,000	7

effect is shown in Table 8.2.

Sawdust was the worst bedding for both total coliforms and *Klebsiella*, while straw produced very high numbers of environmental streptococci on teat skin. This agrees with clinical, on-farm experiences, where *Streptococcus uberis* mastitis is commonly associated with straw yards. It is only damp and badly stored sawdust which has a high coliform count. Provided that the sawdust is stored in the dry (i.e. not allowed to ferment) and is kept dry on the cubicle beds, there is no reason why it should not be used as bedding material. It can be particularly useful when rubber mats and automatic scrapers are in use.

IMPORTANCE OF VENTILATION

The fluid loss from a high-yielding dairy cow consuming large quantities of food is enormous. Approximate figures are:

- 4–5 litres per day from skin and respiratory tract (treble this on a very hot day)
- 20 litres per day in urine
- 30 litres per day in faeces

In addition, high-yielding cows produce large amounts of heat. It is therefore vitally important that buildings are designed to be well-ventilated, in order to remove this humidity and that stocking densities are kept in reasonable proportions, to avoid a build-up of moisture.

Table 8.2 Comparison of growth of mastitis organisms in three different types of bedding (*26*)

	Bacterial counts (geometric means)					
	Sawdust		*Shavings*		*Straw*	
	bedding[1]	*teat*[2]	*bedding*	*teat*	*bedding*	*teat*
Total coliforms	5.2	127	6.6	12	3.1	8
Klebsiella	4.4	11	6.6	2	6.5	1
Streptococci	1.1	38	8.6	717	5.3	2064

[1] Count g/used bedding ($\times 10^6$)
[2] Count from teat swab

Long, narrow, blind-ended straw yards where you can 'feel' the humidity and stale air at the far end are particularly dangerous (see Figure 8.15). Similarly, poorly ventilated cubicle buildings with a low roof, where condensation drips onto both cows and bedding on a cold morning, will predispose to mastitis and respiratory diseases such as infectious bovine rhinotracheitis (IBR). If you can't see the far end of the shed because of condensation and mist, or if condensation is dripping from the roof onto the cows' backs, then ventilation is totally inadequate!

Some suggested ways of ensuring adequate ventilation include:

- an adequate apex outlet at the roof. This is best achieved by having a 23–30.5 cm gap (approx. 9–12″), plus a 15 cm (approx. 6″) 'upstand' on the final sheet of roofing material (see Figure 8.1). A cross-flow of air across the upstand produces an extractor effect. The conventional roofing cowls with a narrow outlet simply do not allow sufficient air flow (Figure 8.2a).
- if a new building is being constructed, turn the roofing sheets upside down and leave a 1.3–1.9 cm (approx. 0.5–0.75″) gap between each run of sheets (Figure 8.2b). This seems to prevent rain from entering and

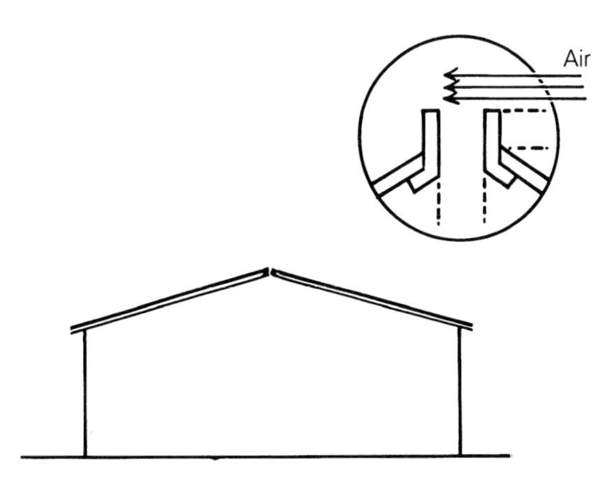

FIGURE 8.1 Leaving a 23–30.5 cm gap at the apex of the roof and adding a 15 cm upstand improves ventilation.

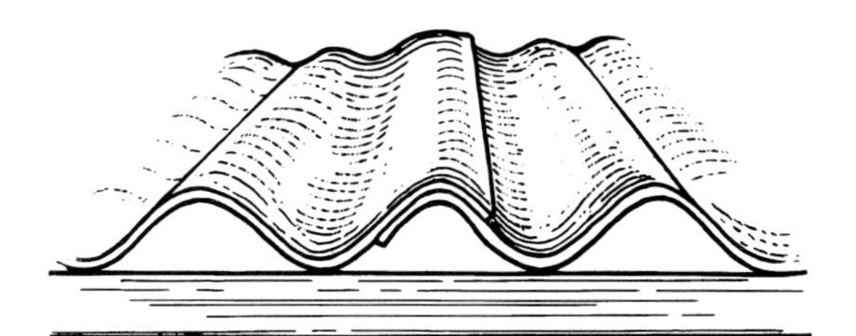

FIGURE 8.2a Conventional roof sheeting in which the edges are faced downwards and adjacent sheets overlap.

FIGURE 8.2b It has been suggested that ventilation can be improved by reversing the sheets so that the edges face upwards and a gap of 1.3–1.9 cm is left between each sheet.

improves ventilation particularly well when the building is full of cattle. It also reduces roofing costs, as fewer sheets are used!

- a similar effect may be obtained in existing buildings by using an angle grinder to cut narrow slots into the top of every 4th–6th ridge of the roofing sheets (Figure 8.3). If this is done close to the apex of the roof it will have a particularly good effect.

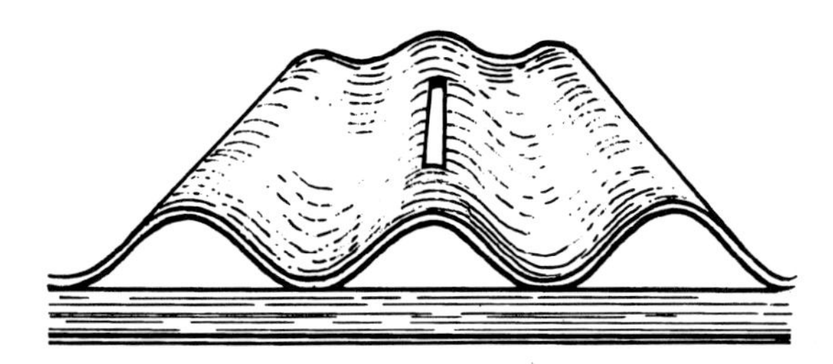

FIGURE 8.3 Ventilation can be improved in existing buildings by cutting slots into the top of every 4th-6th ridge of the roofing sheets.

- clad the sides and gable ends of the buildings with Yorkshire boarding (spaced vertical boarding), leaving a 12.7 cm (approx. 5″) gap between boards.
- avoid multiple-span buildings (Figure 8.4). By far the best air flow is achieved when buildings stand singly. Also avoid buildings of excessive span, e.g. more than 18.3 m wide (approx. 60′).

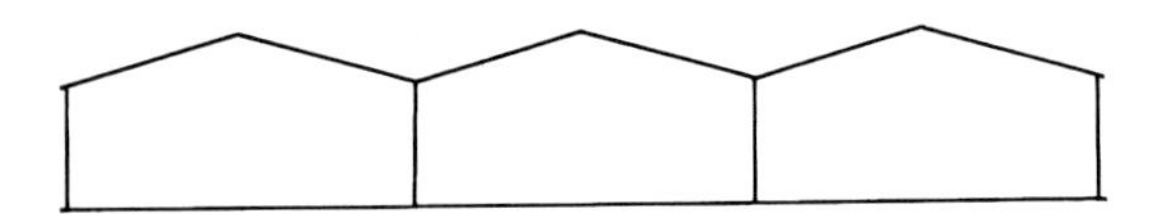

FIGURE 8.4 Avoid multiple span building such as this. The best ventilation is obtained from a single span building which is not excessively wide.

- ensure adequate drainage. Standing water increases the humidity within a building and further predisposes to mastitis. Straw yards with earth or sand floors and cubicle houses with slatted passageways both reduce the amount of standing water.

CUBICLE (FREE-STALL) SYSTEMS

The most important features of cubicle systems are their design and management.

Cubicles should be designed to be comfortable for cows, to be in constant use but to stay reasonably clean. Uncomfortable cubicles will often stay clean, simply because the cows do not use them! Unfortunately, so many cows then lie outside on the concrete that mastitis (and lameness) becomes a problem. Plate 8.4 is a typical example. In this instance a combination of warm weather and uncomfortable cubicles led to a large number of cows lying out, with a resultant increase in mastitis.

PLATE 8.4 A combination of uncomfortable cubicles and warm weather encouraged a large number of cows to lie outside, with a consequent increase in mastitis.

A large amount of the recent work on cubicle comfort has been carried out by John Hughes and the University of Liverpool Lameness Group and part of the following is taken from their ideas.

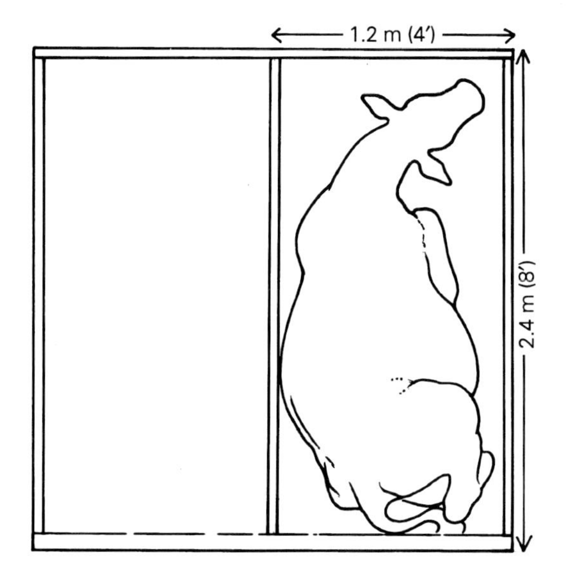

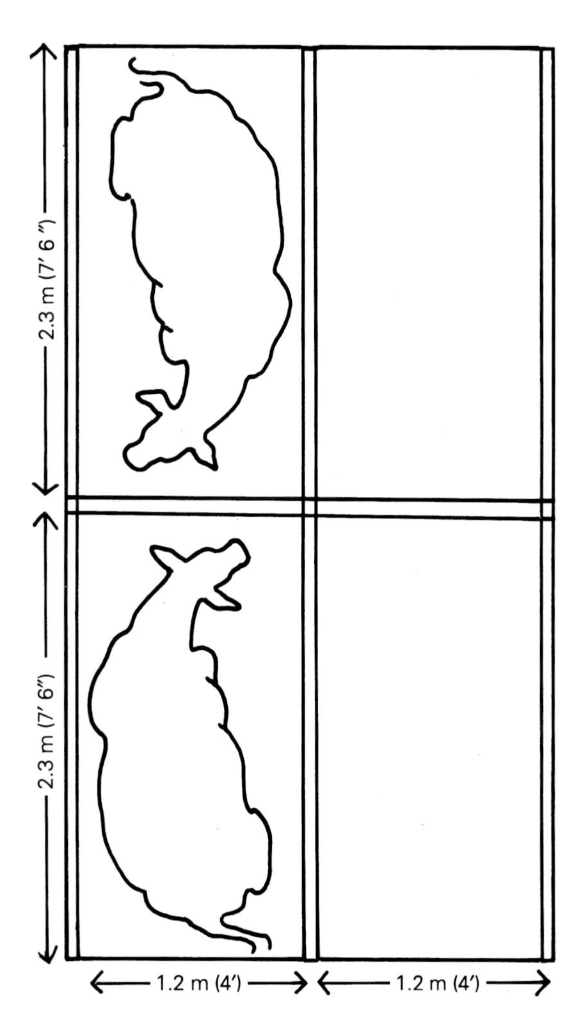

FIGURE 8.5 A single row of cubicles needs to be 2.4 m (8′) long (above). Two facing rows can be 2.3 m (7′ 6″) long (right).

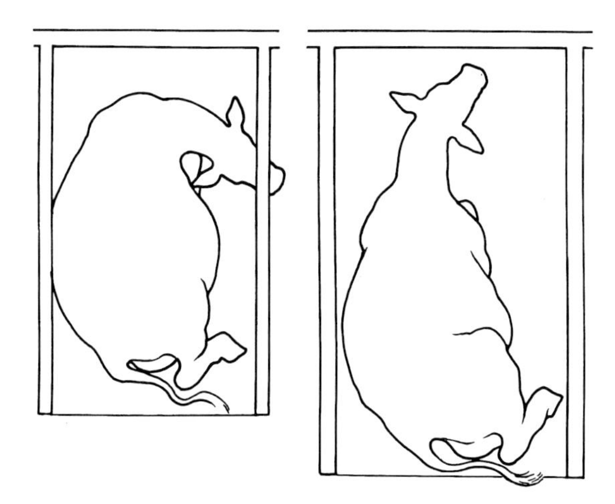

FIGURE 8.6 (Left) A cubicle which is too short forces the cow to sit with her neck excessively flexed, thus impeding rumination. (Right) Space is needed at the front of the cubicle to allow the cow to extend her neck during regurgitation of the cud, and to allow forward lunging when rising to stand.

Size

This must depend on the size of the cow, but for modern large Holsteins 2.3–2.4 m long by 1.2 m wide (7′ 6″–8′ long by 4′ wide) are reasonable dimensions (see Figure 8.5). Length seems to be the most important dimension affecting cubicle acceptance. Where there is a double row of facing cubicles (Figure 8.5 to the right) space sharing at the front makes a 2.3 m length acceptable. The cubicle should be such that a cow can sit in it, with her head held extended forwards to ruminate. If the cubicle is too short, she has to sit with her neck flexed, which makes it difficult to regurgitate the cud (Figure 8.6).

This type of cubicle also encourages cows to stand for excessive periods of time and predisposes to lameness. Cubicles which are too narrow, or which have excessively rigid divisions, can lead to compression of the rumen when the cow is lying down. This can further impede rumination as well as discourage cubicle acceptance. Figure 8.7 shows one such design. Note how the rumen of a large cow would be directly compressed by the cubicle division. By removing both the upright bar AB and the horizontal CD, and replacing them with a length of rope under tension, as in Figure 8.8, the cubicle becomes much more comfortable.

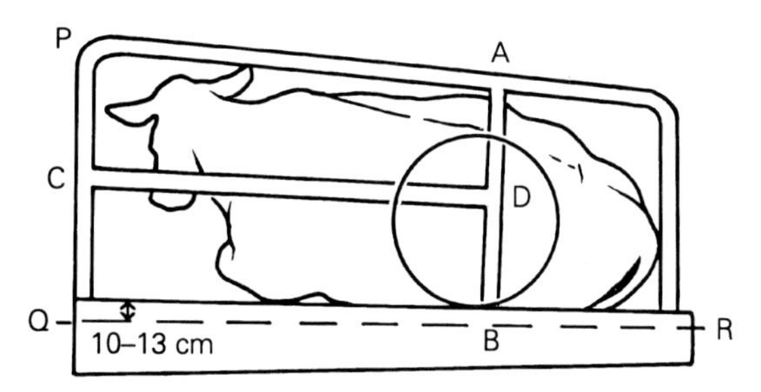

FIGURE 8.7 Cubicles with excessively rigid divisions can be uncomfortable. Pressure is exerted on the rumen (circled).

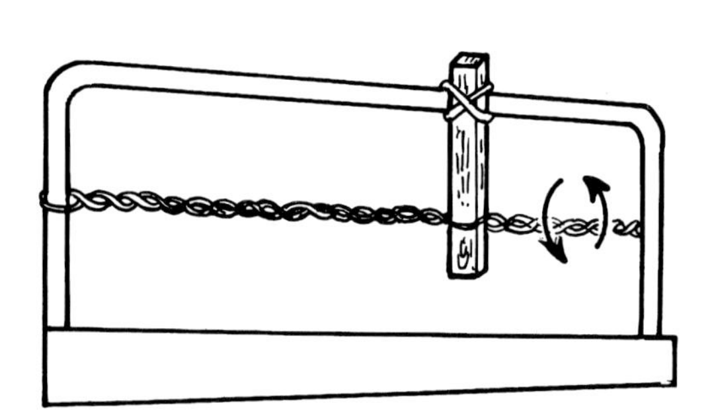

FIGURE 8.8 A flexible division gives greater cow comfort. The lower cubicle rail can be replaced by a rope under tension. A two-stranded rope is brought under tension by rotating a piece of wood (see arrows) fixed between two strands. When the rope is taut, the wood is tied to the top cubicle rail.

Division Height

The height of the division is also important, especially at the front of the cubicle. If height PQ (Figure 8.7) is too short, then the cow has to depress her neck when sitting 'space-sharing' with the adjacent or opposite cubicle. This further reduces comfort and impedes rumination. Clearly the optimum height varies with the size of the cow, but an ideal of 1.125 m (3′ 9″) has been suggested.

Cubicle Length

The problem with long cubicles is that cows may get too far forwards and defaecate on the bed. This can occur either lying or standing. Some animals also occasionally shuffle so far forwards on their knees, finishing up so close to the front wall (as in Figure 8.9) that they either have great difficulty in standing, or are totally unable to rise. Ideally a cow should have at least 1.2 m (4′) of forward lunging space, to enable her to stand easily.

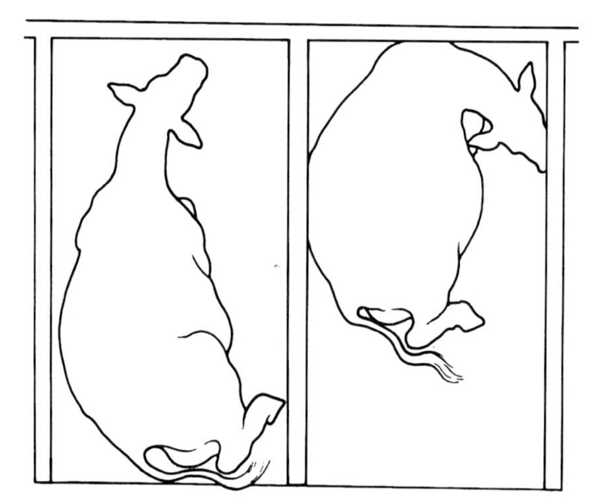

FIGURE 8.9 Some cows move so far forward in the cubicle that they are unable to lunge forward and therefore find it very difficult to stand up.

Neck Rails

Encouraging the cow to remain in the correct position in the cubicle can be achieved by using either brisket boards or neck rails or both. Neck rails can either be attached to the top of the cubicle divisions, or suspended above as shown in Figure 8.10a and b. In either case they should be positioned approximately 30–45 cm (approx. 1′–1′ 6″) from the front of the cubicle, although this varies enormously depending on cubicle length. The suspended rail, positioned to be 7.5–10 cm (approx. 3–4″) below the neck height of a standing animal, is preferable, since rails attached to the cubicle division may be so low as to discourage cubicle acceptance. Both have the same disadvantage, however, in that once she is lying down there is nothing to prevent the cow shuffling into the position shown in Figure 8.9. However, when she stands up, the presence of the rail on her neck will encourage her to reverse to the rear of the cubicle and thereby urinate and defaecate into the passage.

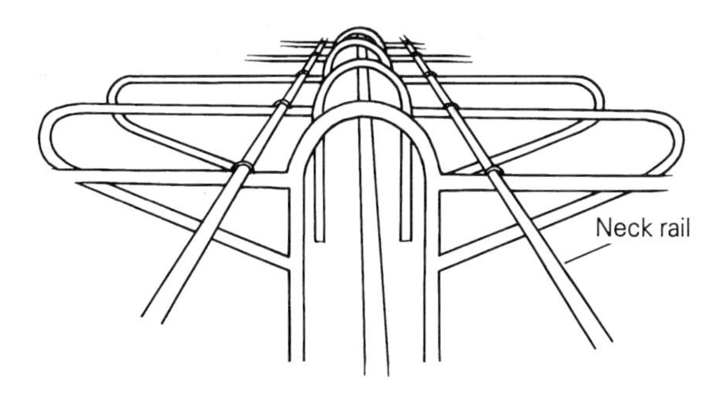

FIGURE 8.10a and b Neck rails may either be fixed to the top of the cubicle divisions (as above) or preferably suspended above the divisions (as shown below).

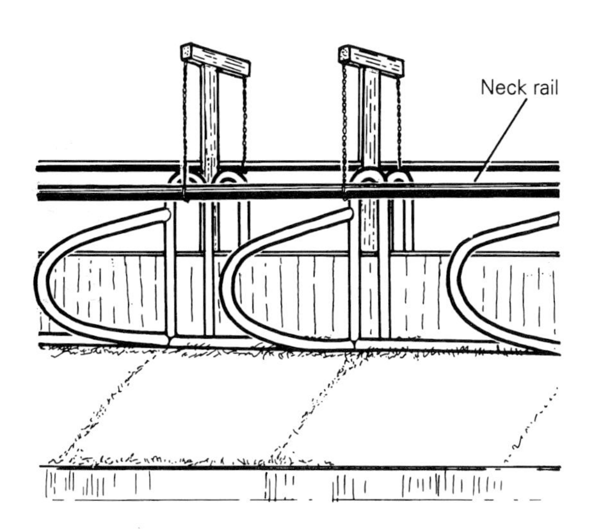

Brisket Boards

Brisket boards sited 1.72 m (5′ 8″) from the back of the cubicle (Figure 8.11) will certainly stop the cow shuffling too far forwards. However, when she stands up she can still easily stand over the front of the board and defaecate onto the rear of the cubicle. A neck rail is therefore needed in conjunction with a brisket board.

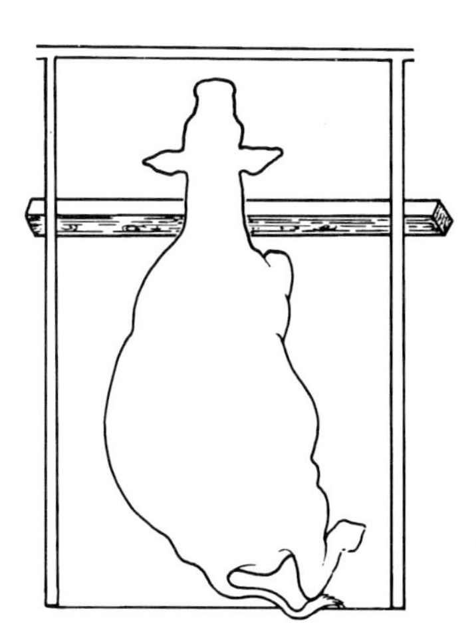

FIGURE 8.11 A brisket board prevents the cow shuffling too far forwards. It should be angled forward towards the front of the cubicle to reduce knee damage.

A novel idea, which should overcome the problems encountered with neck rails and brisket boards, has recently been introduced (Hughes). It involves inserting

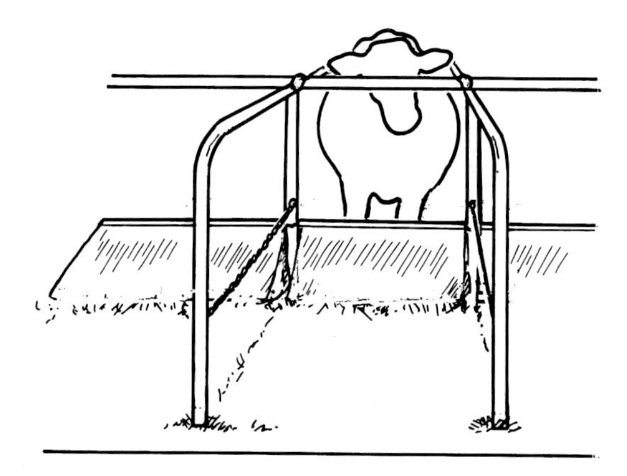

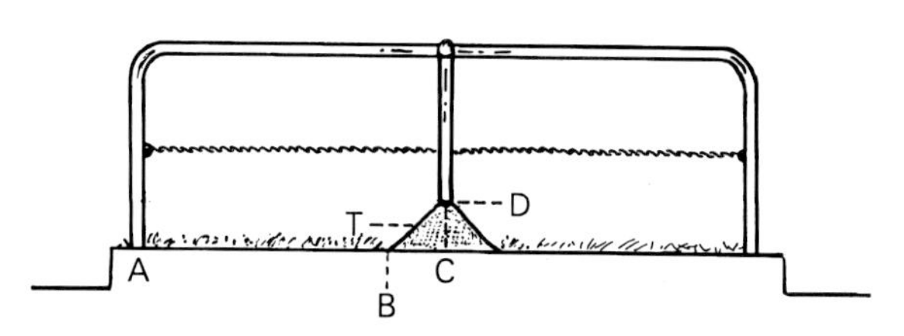

FIGURE 8.12 A pyramid of concrete between two facing rows of cubicles (or a triangle against a wall) prevents cows moving too far forward, while at the same time allowing sufficient space for cudding and lunging forward to stand.

a long, pyramidal concrete shape, between two facing rows of cubicles (Figure 8.12 and Plate 8.5), or a single triangle at the front of a single row. Hughes suggests 1.72 m (5′ 8″) for the distance AB from the rear kerb of the cubicle and 0.38 m (15″) for the vertical height CD of the concrete. When seated, the cow can no longer go too far forwards in the cubicle and yet the height CD means that she can extend her neck over the top of the pyramid in facing rows of cubicles. When rising to stand she may place one foot on T, on the slope of the concrete, to push herself upright, but when fully standing she will have to keep her feet back behind B and hence will defaecate into the dunging channel. A neck rail is often no longer necessary and this may further increase cubicle acceptance.

PLATE 8.5 A 0.38 m high concrete pyramid between two facing rows of cubicles prevents cows going too far forward and yet still allows rumination.

Cubicle Base

Limestone, earth, sand and concrete have all been used for cubicle bases. The first three all suffer the disadvantage that they gradually become eroded to form pits, which at the rear of the cubicle can become filled with damp, soiled bedding which will then represent a source of mastitis infection. Plate 8.6 shows a typical example.

PLATE 8.6 Badly soiled cubicle beds represent a mastitis risk.

Figure 8.13 shows the three major points of contact when a cow is lying in a cubicle. These are the positions of the two knees (A and B) and either the right hock (as in Figure 8.13) or the left hock, depending on which side the cow is lying (C). These three contact points are clearly recognisable in many cubicles: look for the three areas where straw bedding has been scuffed away, often exposing bare concrete. Unless concrete is used, cows continually lifting themselves with their hind feet can eventually erode a depression at the back of the cubicle, and this can then become soiled with faeces and wet bedding, as in Plate 8.6.

If limestone is used, then it **must** be covered with adequate bedding, otherwise stones at A and B (Figure 8.13) will lead to severe discomfort and knee trauma when the cow tries to stand up. Similarly stones, or an excessively sharp or rough kerb or heelstone can lead to trauma and swellings of the hocks. A typical example is shown in Plate 8.7. The swellings initially consist of bruised skin, sometimes with a quantity of fluid accumulating in the hock bursa. (A bursa is a small 'shock absorber' pouch which acts to protect a protruding portion of bone and allows skin, muscles, tendons etc. to glide over the bony surface.) Only in the very late stages, when the skin is broken, would the swellings in Plate 8.7 become infected.

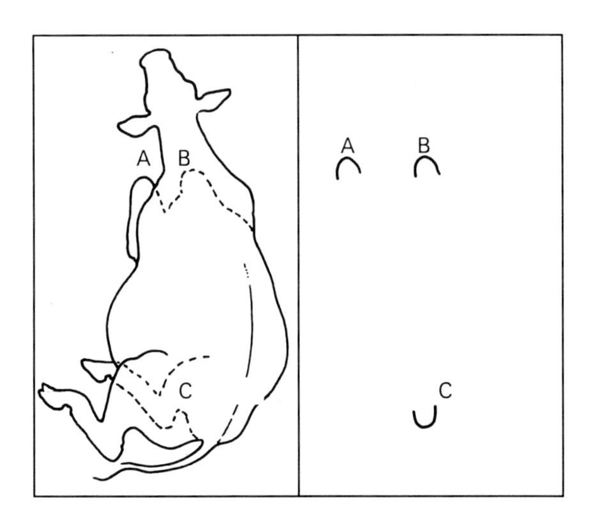

FIGURE 8.13 A, B and C are the major contact points between the cow and the cubicle base and must be well bedded.

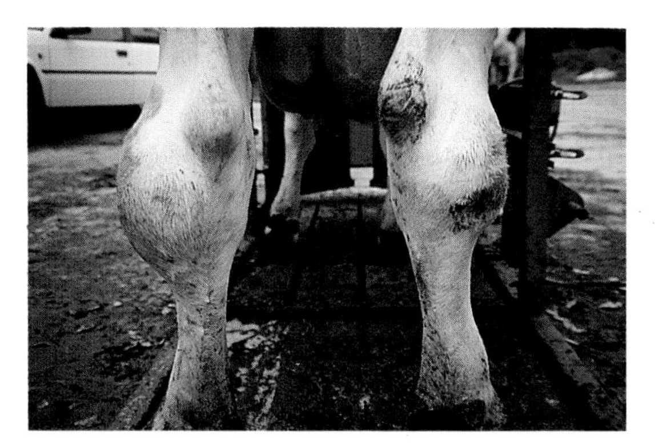

PLATE 8.7 Gross hock swellings, the result of lying on a hard surface. The hair-loss and scab on the skin surface show the major points of abrasion.

Many farms now have concrete bases in their cubicles. Although these are certainly easier to keep clean, they can be hard and uncomfortable and this may lead to cubicle rejection.

If animals are accustomed to being in cubicles having a lip, they often touch the end of the lip with their toe before stepping down into the dunging passage. Removal of this lip (by concreting the cubicle base) can induce apprehension in some animals because they do not know when and where to step down and this too can lead to cubicle rejection. Similarly, cubicles with an excessively high kerb (e.g. greater than 12.5–15 cm or 5–6″) may be rejected, as heifers particularly are often nervous about reversing down off a high step.

The slope of the cubicle floor is also important and should be 10–13 cm (4–5″) from front to rear, that is from Q to R in Figure 8.7. A cow much prefers to lie uphill. A level or, even worse, downward-sloping cubicle could lead to rejection by some cows. They may then lie outside on dirty concrete, thus predisposing to mastitis and lameness. In addition, any urine which splashes onto the rear of the cubicle will run forwards, soiling the bed and further increasing the risk of mastitis. Cubicle lips should be discouraged as they tend to hold fluids, urine and milk and impede drainage along the cubicle base.

Cubicle Bedding

Although different types of bedding support different types of mastitic organisms (see pages 39 and 105), probably the most important factor is to ensure that cubicles are kept **clean and dry**, i.e. cubicle management must be good. Sand, straw, sawdust, shavings, shredded paper and ground limestone have all been used for bedding, with straw being the most common in the United Kingdom. Comfortable cubicles are also important to reduce the incidence of lameness. Rubber mats are good for cows, but they **must** be kept dry. If the surface becomes wet, then this predisposes to mastitis. Natural rubber is said to become slippery when in contact with the oils from milk and skin and some say that synthetic rubber is preferable. Cow mattresses (see Plate 8.8), filled with shredded rubber, have also been used to increase comfort.

Mats should extend to the rear lip of the cubicle, otherwise the hock may become damaged by lying on the edge of the mat (Plate 8.7). Cubicle bedding is still required to prevent hock sores, even with mats although reduced amounts are as acceptable. In this instance the bedding seems to act as a 'lubricant', reducing friction between hock and mat.

PLATE 8.8 Cow mattresses: canvas sacks filled with rubber chippings.

When available, ample quantities of straw are ideal. Figures of 350 kg per cow per winter have been quoted, but this varies enormously with the depth of straw provided in the bed and whether or not mats are present. 350 kg is probably a minimum figure in the absence of mats. Shredded paper has been used as bedding, but has not become popular in the United Kingdom. It is not particularly absorbent and when wet it tends to become matted and solid. It also tends to stick to cows' flanks and looks untidy.

Management

Ideally, all soiled and damp areas should be scraped from the backs of the cubicles at least twice daily (and preferably every time the herdsman walks past) and fresh bedding added once daily. If straw usage is liberal, it may be sufficient to bed the cubicles twice weekly, or simply to scrape fresh straw from the front to the rear of the cubicle every day, as required. A small quantity of hydrated lime ($Ca(OH)_2$) or even ground limestone ($CaCO_3$) sprinkled onto the rear of the cubicles once a week (as in Plate 8.9) also helps to keep the bed dry, as lime absorbs moisture. Lime should be applied and then covered with fresh bedding (e.g. straw or sawdust). This prevents direct and excessive contact of lime with teats, which could otherwise lead to cracking. Do **not** use quicklime (CaO) as this will produce severe teat burns.

PLATE 8.9 Hydrated lime both dries and disinfects cubicle beds. However, do not use an excess as it can also produce drying and cracking of the teats.

The importance of regularly renewing the bedding is shown in Figure 8.14. When sawdust was added to cubicles on a weekly basis (A), coliform levels were

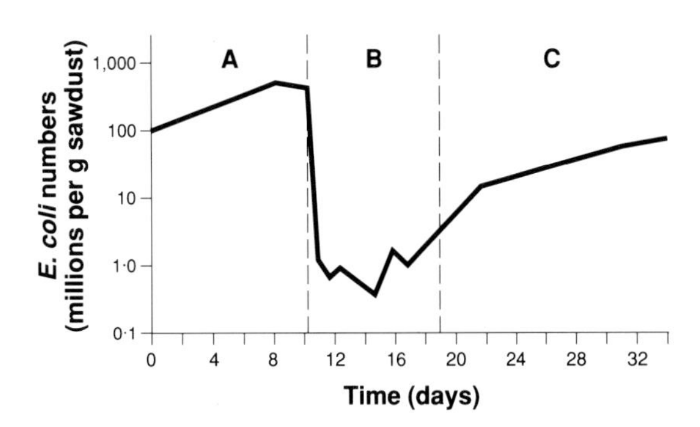

FIGURE 8.14 The importance of regular renewal of cubicle bedding. E. coli *numbers were high when fresh sawdust was added weekly (A), fell rapidly when daily bedding was introduced (B), but soon deteriorated on return to weekly bedding (C). (27)*

quite high. Levels declined when daily bedding was carried out (B), but the situation soon deteriorated when there was a return to the original system of weekly bedding (C).

Cubicle passages should be scraped at each milking and ideally this should be carried out **before** the cows return to the cubicles. This keeps the teats as clean as possible during the first critical 20–30 minutes after milking, when the cow is more susceptible to mastitis, because the teat sphincter has not fully closed (see page 18). It also reduces the amount of faeces carried back onto the cubicle beds by soiled feet.

STRAW YARDS

Straw (loose) yards (plate 8.2) are certainly good for cow comfort and, if given the choice, cows would opt for yards rather than cubicles. However, they are not without problems. Straw usage is much higher (3–6 times more) than in cubicles and hence both bedding and labour costs are higher. Stocking densities tend to be lower, because you can keep fewer cows in a straw yard than if the same building were used for cubicles. Straw yards are also often associated with an increased mastitis risk, especially if they are not adequately managed. However, there is generally a lower incidence of lameness.

Bedding

Yards should be bedded daily, preferably during morning milking and, as with cubicles:

- cows should be kept off the yard for 30 minutes after milking.
- passageways leading back to the yards should be scraped clean before the cows walk along them.

The straw used for bedding must be clean, dry and free from fungi and moulds. Outbreaks of mastitis can occur in association with the use of damp and mouldy straw, even if liberal quantities are used. Mastitis caused by yeasts and moulds is a particular problem, because of its poor response to treatment (see page 41).

Straw beds tend to heat up. Fortunately the anaerobic fermentation which occurs in the compacted lower levels does not support the growth of *E. coli.* Surface temperatures in normal yards are likely to be around 40°C. There is some suggestion that the use of excessive quantities of straw can lead to overheating and that this can increase coliform counts. Yards should be cleaned out frequently, **at least every six weeks**. If left longer, there is an increased risk of mastitis. Some farms clean out as frequently as every 1–3 weeks and suggest that this uses less straw.

Hydrated lime should not be used. Lime tends to prevent normal fermentation in the lower level of the bed and can result in such an uneven surface to the yard that cows find it difficult to walk around. There is some evidence that a hardcore base is better than concrete, in that it permits better drainage. However, this is only likely to be of major importance in yards where the concrete base is flat and poorly drained. If the straw squelches when you stand on the bed, then it's too wet!

Yard Design

One of the most important aspects of straw yards in relation to mastitis is the design of the yard. Long, narrow yards (as in Figure 8.15, on the left) are more easily soiled, because cows have to walk across a greater distance to get to the rear. The positioning of the water troughs in the example shown is also very poor as the only access to them is through the bedded area.

The design shown on the right is far better. Access to the water troughs (W) should only be from the feeding passage (P) and hence this avoids excessive soiling of the bedded area (and it is less important when the water trough overflows!)

Opinions differ on the value of the barrier BC. By restricting access, areas AB and CD become more soiled, but the area behind BC stays cleaner and cows prefer to lie against a wall if possible. Some systems have a continuous step approximately 30 cm high (12″) running from A to D. This helps to retain the straw bedding and gives full access to the yard. The depth of the yard (AE) should be at least 7.3 m and, preferably, no greater than 9.1 m (24–30′), with a feed passage (P) of 3.5 m (approx. 12′) and a food trough (or floor area) (F) of 0.76 m (2′ 6″). Scraping the passage (P) twice daily further reduces faecal soiling of the bed.

Ventilation is equally important in straw yards as in cubicle houses and cow sheds, and can be improved by providing roof insulation. However, this is rarely done because of cost. The ideal humidity for dairy cows is around 70%, whereas many buildings reach 85% or higher in the United Kingdom during winter. High humidity also increases the heat stress in cows kept in hotter climates and should be avoided if possible.

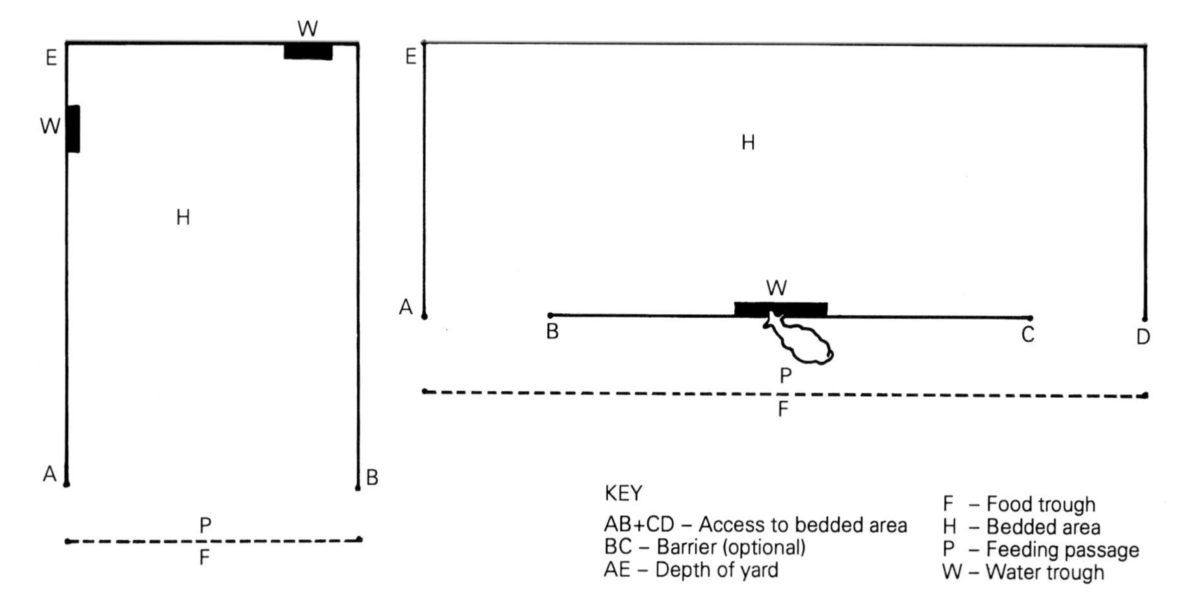

FIGURE 8.15a and b Design of straw yards: long, narrow poorly ventilated yards with badly placed water troughs should be avoided (left). A more useful design is shown on the right.

SAND YARDS

In hotter climates and desert regions cows are housed on sand corrals or yards and may have access to cubicles or a slatted area under cover. Shaded areas are vital and in the absence of specific shade cows tend to congregate and shelter along the edges of buildings, as seen in Plate 8.10. More commonly, tall constructions providing shade are erected, and cows will lie in their shadows. The dimensions and siting of such shaded areas are vital. Ideally they should provide shade over different parts of the sand yard throughout the day, so that all areas under shadow are **also** exposed to the drying influence of the sun at least once each day. Plate 8.11 shows cows lying under a sun shelter.

PLATE 8.10 In the absence of specific shelters, cows congregate along the side of buildings to obtain shade.

PLATE 8.11 Artificial shade from the sun in hotter climates. Note how the cows specifically lie in the areas under shadow.

During the dry season corrals should be cleaned out every 6–8 weeks. The top surface of the sand is scraped off and removed. In the Middle East soiled sand (sand and dry faeces) is a valuable commodity for horticulture. Fresh sand is brought in to top up the yard (Plate 8.12).

PLATE 8.12 Yards should be cleaned and fresh sand provided every 6–8 weeks.

Sand provides good drainage and the heat of the sun dries the faecal pats, which are then broken up each day by a tractor and scraper (Plate 8.13).

During wet weather sand yards become very muddy (see Plate 8.14) and cleaning teats prior to milking becomes a major task. The risk of environmental mastitis increases substantially. If possible, cows should be kept in cubicle areas, preferably with fans as a cooling system, until the yards dry out.

PLATE 8.13 Sand yards need to be scraped daily to break up dried faecal pats.

PLATE 8.14 Under wet conditions, open sand yards become a real problem.

GENERAL CONSIDERATIONS

There are certain points of general management which are applicable to **all** housing systems.

Avoid High Stocking Densities

Tightly packed cows create high humidity and are often under stress, especially younger heifers. Whenever possible, a large loafing area should be provided (Plate 8.15). In many parts of the world this need not be totally under cover, since cows are prepared to go outside in quite low temperatures, as long as it is neither raining heavily nor extremely windy. In hotter climates, loafing areas can be used at night. The provision of adequate loafing (and feeding) areas also aids in heat detection and helps to reduce the incidence of lameness. This is because

PLATE 8.15 In temperate climates, access to clean, open loafing and feeding areas helps in mastitis control, reduces lameness and improves heat detection.

PLATE 8.16 Waste silage left beside troughs encourages cows to lie outside and can predispose to environmental mastitis.

cows which have enough space to walk around are likely to suffer less damage to their feet than cows which stand still for long periods of time. In addition, if they are able to move away from other cows not only are they more likely to show more signs of oestrus (heat) behaviour, but it will be easier to identify such cows.

Clear Away Waste Food

Waste silage or other food left lying beside the trough encourages cows to lie outside (Plate 8.16). It can also be a good culture medium for environmental mastitis bacteria (see page 29), particularly *E. coli, Bacillus licheniformis*, and *Bacillus cereus*, and by contaminating the teats can produce high TBCs. Areas around the feed troughs should therefore be cleaned regularly (see Plate 8.15).

Handle Cows Gently

There is plenty of evidence that stressed cows are more prone to infections and this includes mastitis. If rushed along roads and through doorways they may injure or excessively soil their teats. If forcibly driven into the milking parlour, the cows' let-down is likely to be inhibited, with the consequences of increased residual milk and depressed yields (see page 90). The difference between quietly and roughly handled cows very soon becomes apparent by their reaction to visitors.

Avoid Draughts

Chilling of the udder may reduce the immune response and hence the cow's ability to counteract infections which have penetrated the teat canal. Chilling of teats will undoubtedly lead to cracking and chapping and this further predisposes to mastitis. This is particularly important after milking when cows are left to stand for 20–30 minutes to allow the teat canal to close. They must not be left standing in exposed yards or draughty passageways, especially when the teats are still wet with teat dip.

Establish a Post-calving Group

Larger herds in Arizona, California and the Middle East already have an immediate post-calving or maternity group of cows, which are retained in a more 'gentle' environment than the remainder of the herd. As cows are at their most susceptible to mastitis and in particular to acute coliform infections at this stage, it must be a logical step. A small, but increasing number of farmers in the United Kingdom are now following this example, by keeping their freshly calved cows in straw yards for the first 4–6 weeks after calving, before introducing them to the main herd and to cubicles. It has been found that this can increase yields and decrease lameness. Perhaps surprisingly, cubicle acceptance is improved when the change from straw yards to cubicles occurs. Of course, it is vitally important that this group should be retained in a clean and well-bedded yard, at a low stocking density, otherwise mastitis problems could be exacerbated. Pre-dipping (page 99) should definitely be carried out in this group, if it is not already in routine use.

CHAPTER NINE

Somatic Cell Count

The somatic cell count is the number of cells present in milk ('body' cells as distinguished from invading bacterial cells). It is used as one indicator of udder infection. Somatic cells are made up of a combination of white blood cells and epithelial cells. White blood cells enter milk in response to inflammation which may occur due to disease (see page 22) or occasionally injury. Epithelial cells are shed from the lining of the udder tissue. White blood cells make up nearly all (about 98–99%) of the somatic cells.

The somatic cell count is quite a crude measurement and there are a variety of factors that will affect the result. In general, it is the **contagious** mastitis organisms that are responsible for high cell counts as they make up the majority of subclinical infections. This is because the body continues to send in large numbers of white cells while attempting to remove this subclinical infection.

The somatic cell count, or SCC for short, is measured in thousands of cells per ml of milk. The results are normally expressed in thousands to the farmer, e.g. a count of 250 refers to 250,000 cells per ml of milk. It is impossible to have a cell count of zero.

In 1994, the annual average herd cell count in England and Wales was 260,000 and 75% of all dairy herds had a cell count under 400,000. This compares favourably with 1981 when under 40% of herds had a cell count under 400,000 (see Figure 9.1). This trend will continue as the dairy companies impose higher financial penalties for high cell count milk.

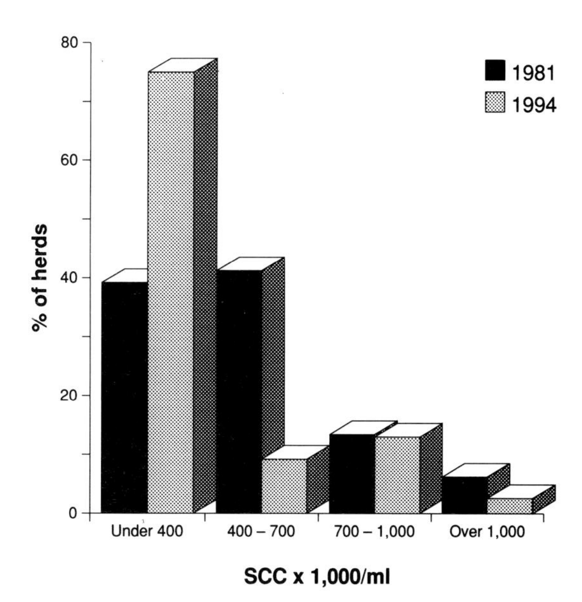

FIGURE 9.1 Cell count breakdown of herds in England and Wales, 1981 and 1994. (28)

There is no reason why any dairy herd should not have a mean annual rolling herd cell count under 150,000. This means a low level of subclinical infection and minimal damage to the milk-producing tissues thereby maximising milk yield, and ensuring the production of quality milk that will attract a premium price.

WHY CELL COUNTS ARE IMPORTANT

Financial Penalties

Nearly every country now has some system of financial penalty that is imposed if the cell count or TBC (total bacteria count, see Chapter 10) of bulk milk rises above a certain threshold. This is intended to ensure that the milk produced is of the highest quality. Farmers who do not meet these production standards are penalised according to the quality of their milk.

In 1994 all dairy companies in the UK penalised farmers with high cell counts. Some companies penalised producers with counts over 100,000 using a sliding scale of increasing penalties for higher counts.

The European Union has laid down rules for the production and market use of raw and heat treated milk as well as other dairy products in its Health and Hygiene Directive 92/46/EEC. In the United Kingdom, the first stage of the directive came into effect in January 1994 when milk for liquid consumption had to have a cell count of no more than 400,000 and milk for manufacturing no more than 500,000.

The final stage of this directive will be implemented on 1 July 1997. This will make it illegal to use milk with a cell count over 400,000 for both liquid consumption and manufacturing purposes. In effect, the introduction of financial penalties for high cell count milk by the various dairy companies should ensure that nearly all dairy producers meet the EU standards well before 1997.

Reduction in Milk Yield

Most farmers are well aware that as the herd cell count rises there is a corresponding drop in milk yield. This occurs as a result of damage to the milk-producing tissue caused by mastitis bacteria and the toxins that they produce.

Canadian work has shown that milk yield drops by 2.5% for every increase in cell count of 100,000 above a base figure of 200,000. This is shown in Figure 9.2. For example, a herd with a count of 500,000 can be expected to have a 7.5% loss in yield due to subclinical mastitis. In herds with good mastitis management, a cell count of 200,000 can be maintained with ease and so this figure was assumed as the baseline at which there are insignificant production losses.

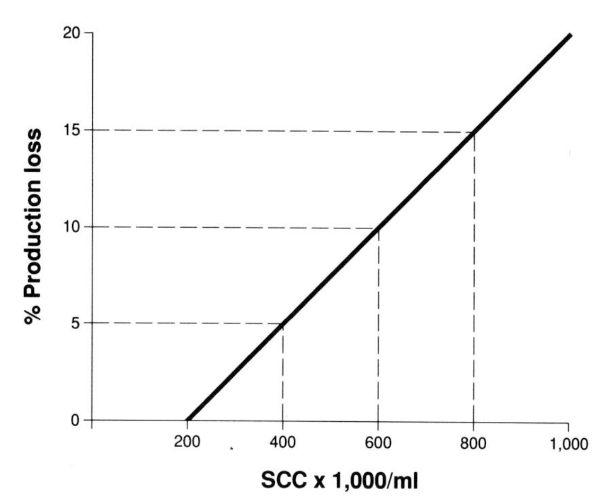

FIGURE 9.2 Effect of herd cell count on milk production: milk yield drops by 2.5% for every increase in cell count of 100,000 above a base figure of 200,000. (29)

The Suitability of Milk for Manufacturing or Liquid Milk Consumption

The final and most important concern about high cell counts is the acceptability of milk to retailers and the manufacturing industry. It must be remembered that the quality of milk is only as good as when it leaves the farm. Poor quality milk always remains poor quality milk.

High cell count milk has a high level of the undesirable enzymes lipase and plasmin (see page 14). Lipase breaks down fat, produces a rancid flavour, inhibits yoghurt starter cultures and will reduce the shelf life of milk. Plasmin decreases the amount of casein in milk and will reduce the yield of cheese. It continues to act even in cold storage conditions and after pasteurisation.

MEASUREMENT OF CELL COUNTS

Fossomatic and Coulter Counters

The dairy industry can use one of two forms of automatic measurement: the Fossomatic counter or the Coulter counter. Cell counts in the UK are measured using the Fossomatic method which has now superseded the Coulter counter in virtually every country throughout the world. There can be a considerable level of variation in the measurement of milk samples using either of these methods, with a difference of up to ± 15%. Bulk tank and individual cow samples are all tested using automatic methods.

Californian Mastitis Test (CMT)

This is a simple test that is useful in detecting subclinical mastitis by crudely estimating the cell count of milk. The CMT test does not give a numerical result, but rather an indication whether the count is high or low. Any result over a trace reaction is regarded as suspicious.

The benefits of the CMT:

- a cheap test.
- can be carried out by the milker during milking.
- results available immediately.
- gives an indication of the level of infection of each **quarter** as compared to an individual cow cell count that only gives an overall **udder** result.

The test is carried out in the following manner (see Figure 9.3):

- foremilk is discarded.
- one or two squirts of milk from each quarter are drawn into the paddle dish.
- the paddle is tilted so that most of this milk is discarded.
- an equal volume of reagent is added to the milk.
- this solution is then mixed and examined for the presence of a gel reaction. The plate must be rinsed before going on to the next cow.

The results can be scored into five categories: negative where the milk and reagent remain watery, to the highest cell count where the milk and reagent mixture almost solidifies. This is determined according to the gel reaction.

Whiteside Test

The whiteside test is a laboratory test (now superseded by the CMT test) in which somatic cells react with the chemical, sodium hydroxide. Coagulation occurs when counts exceed 500,000.

Direct Microscopic Stain

The direct microscopic somatic cell count is rarely carried out now as it is slow, tedious and not very accurate. It has been superseded by electronic tests. A volume of milk is spread over a calibrated slide, stained and the number of cells counted under the microscope.

FACTORS THAT AFFECT SOMATIC CELL COUNTS

Mastitis

Mastitis is by far the most important factor that causes increased cell counts. When mastitis organisms enter the udder, the defence mechanisms send vast numbers of white cells into the milk to try and kill bacteria (see page 22). If the infection is eliminated, then the cell count will fall back to its normal level. If the white cells are unable to remove the organisms, then a subclinical infection is established. White cells are then continually secreted into milk, leading to a raised cell count.

Foremilk is discarded and one or two squirts of milk are drawn from each quarter into a paddle dish

Excess milk is discarded

An equal volume of CMT reagent is added to the milk

The milk and the reagent are mixed

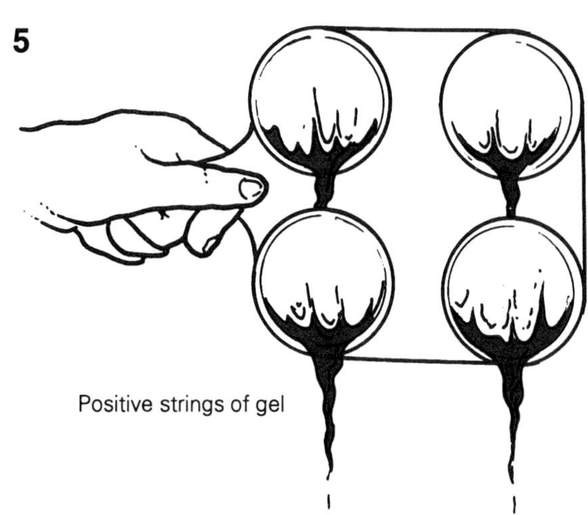

Solutions are examined for the presence of a 'gel' or 'slime' reaction: gelatinous 'strings' indicate a high cell count quarter

FIGURE 9.3 How to carry out the Californian Mastitis Test (CMT).

Type of Mastitis Organism

Contagious bacteria (see page 29) are much more likely to produce subclinical infections and therefore high cell counts than environmental organisms. Infections caused by environmental organisms tend to be rapidly eliminated and the cell count normally only rises during the period of mastitis.

Different bacteria can produce different immune responses in the body. In addition, the same organism may produce differing responses in the same animal.

In the case of acute *E. coli* infections there is often a huge variation in response, as discussed on page 36. When the immune system responds well, there will be a massive increase in the number of white cells, for example up to 100,000,000 per ml within four hours of *E. coli* invading the udder. In other instances, particularly in early lactation cows, the defence mechanism does not react, there is no increase in cell count and the cow will die no matter how she is treated. This is because the organism is free to multiply and produce toxins with no resistance from the cow.

For some organisms there is a relationship between cell count and the level of infection in the udder. For example, severe *Streptococcus agalactiae* infections may produce counts of up to 12,000,000 per ml in infected quarters, and the count correlates well to the level of infection. Other mastitis bacteria, particularly *Staphylococcus aureus*, produce a much more variable response as shown in Table 4.3. There is **no** method of identifying the mastitis organism from the cell count of an individual quarter.

Age

Older cows tend to have higher cell counts. This is due to a variety of reasons. Chronic infections, especially *Staphylococcus aureus*, will be more common simply due to the increased period of exposure of the udder over previous lactations. The teat canals may be damaged allowing easier access for bacteria to enter the mammary gland. Finally, the immune response from older animals may be poorer resulting in less bacterial elimination.

Figure 9.4 shows the distribution of cell counts in a herd quite badly infected with contagious mastitis organisms. This herd was divided into three groups: first lactation heifers, cows in lactation numbers two to four, and older cows in their fifth and subsequent lactation. Only 11% of the heifers had cell counts over 1,000,000 compared to 21% of the middle group, and 46% of older cows in lactation five and upwards. Freshly calved heifers tend to have cell counts in the range of 20,000 to 100,000 and in

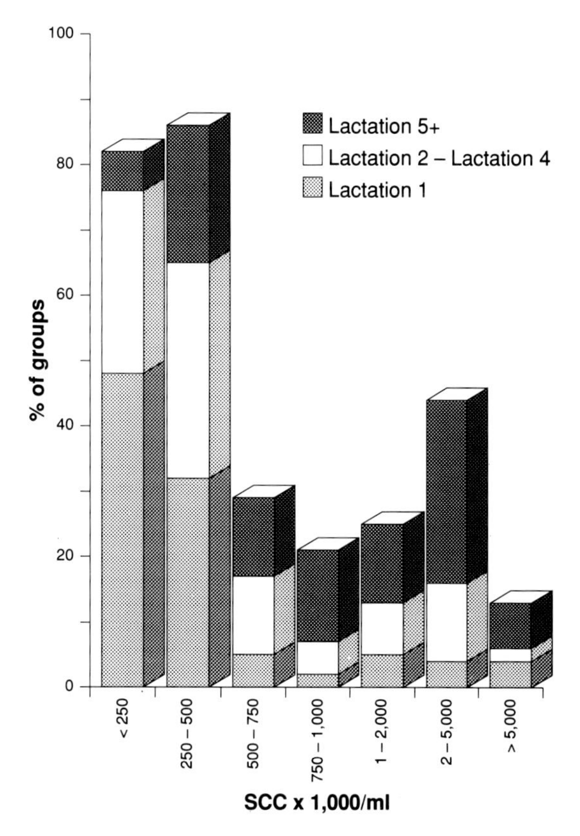

FIGURE 9.4 The distribution of cell counts (SCCs) by lactation in a herd infected with contagious mastitis organisms.

the absence of mastitis would remain at this level.

Stage of Lactation

Cell counts are often high in the first seven to ten days after calving although this may not occur in every cow. Towards the end of lactation, as the amount of milk produced reduces, cell counts rise in animals that have subclinical mastitis. For example a cow producing 10,000,000 cells per day in 20 litres of milk will have a cell count of 500,000. If the same cow was only producing 5 litres, the cell count may increase to 2,000,000. This is due to a concentration effect and is very marked in animals yielding less than 5 litres of milk per day. Cell counts in cows that are free from subclinical infection do not alter significantly in late lactation.

Diurnal and Seasonal Variations

Cell counts tend to be higher in the afternoon milking than the morning milking. This is partly due to a shorter milking interval and lower milk yield resulting in a concentration effect. This can be seen in a herd which had separate tanks for morning and afternoon milk (see Table 9.1). The results for a month were averaged out and show that the afternoon milk had a significantly higher cell count.

Table 9.1

Milking	*Average cell count*	*Variation*
Morning	147,000	±60,000
Afternoon	221,000	±70,000

Counts tend to be higher in summer than in the winter but the reason for this is not clearly understood.

In seasonally calving herds, there may be a rise in cell count when most cows are towards the end of lactation. Figure 9.5 shows the monthly and the annual average cell count for an autumn calving herd over a four year period. The monthly results fluctuate depending on the time of the year. In the summer, when most cows are in late lactation, the monthly cell counts rise, but fall back in the winter. The annual rolling mean removes seasonal factors and so gives a truer long term picture of what is actually occurring on the farm.

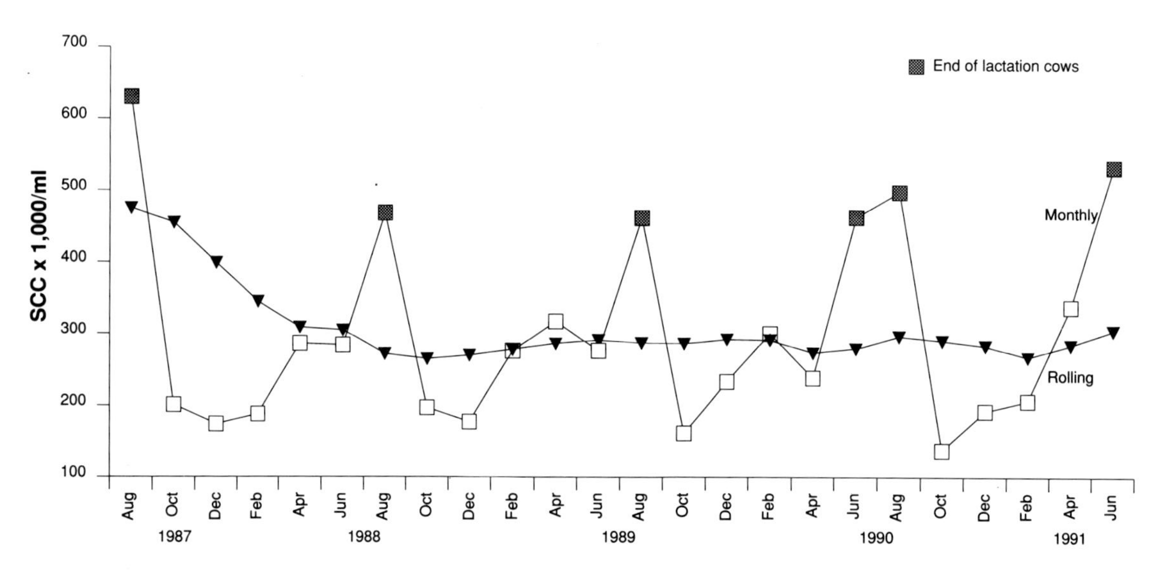

FIGURE 9.5 In seasonally calving herds there may be a rise in cell count when most cows are towards the end of lactation. This is clearly shown in an autumn-calving herd over a four-year period.

Stress

Any event that causes stress such as oestrus (bulling), sickness or events like tuberculin testing may affect the cell count. In addition to increasing the number of white cells in the blood there is frequently a reduction in milk yield and this causes a further concentration effect.

Milking Frequency

Many farmers reduce the frequency of milking to once daily or even every other day before drying off. Research work shows that cows milked intermittently towards the end of lactation will have dramatically **increased** cell counts, even in the absence of subclinical infection.

In tests the average cell count for a group of non-infected cows yielding over 5 litres per day was 237,000. When these cows were not milked for two days the cell count increased to 540,000. A further delay in drying off, and stopping milking for an extra four days increased the cell count to 7,600,000, with some of the cows having counts as high as 15,000,000.

These results clearly show that cows should be dried off **abruptly** and before yields drop below 5 litres a day.

Day to Day Variations, and Management Factors

Cell counts vary from day to day. This is due to a variety of all the factors listed previously together with management factors such as nutrition, calving patterns, sources of replacements and milking machine function. Research work has established that as the level of vacuum reserve in a milking plant decreases the herd cell count increases.

HERD SOMATIC CELL COUNTS

The factors listed above explain the variation in **individual** cow cell counts. Within a herd, many of these variations are averaged out. By far the greatest influence on **herd** cell count is the level of **subclinical** mastitis. As this rises, so does the cell count. A herd with a cell count under 200,000 will have little contagious mastitis present compared to a herd with a count over 500,000 which will have a serious problem. However, herd cell counts do not necessarily correspond to the number of clinical cases, since this could be due to a high level of environmental mastitis that will have little effect on cell count.

Farmers may receive a variety of different sorts of cell count results:

- individual tank results.
- mean monthly figures.
- three month rolling means.
- annual mean counts.

These may all give different results, as shown in Figure 9.6, and trends that can be misleading. Specific results will vary greatly depending on what is happening within the herd on that date. The more samples measured, the greater the amount of variation. The mean annual figure gives a good indication of the overall herd trend as it eliminates the effect of any seasonal variations.

In herds with rising cell counts, two or three sets of low results **may** suggest that the problem has disappeared. In some situations this may be the case, as the offending cow or cows may have been dried off or sold. In the majority of cases however, it is just a temporary fall that will rise again. Figure 9.6 shows the individual, monthly, three monthly and annual rolling mean cell counts for a 150 cow herd over a period of 13 months. It can be seen that there is a high individual bulk result marked 'A' at the beginning of May. The next three individual results are lower, and many farmers may consider that the problem has disappeared. However, it can be seen that over the following three months, all the cell count parameters increase indicating that this was not a one-off incident. It is essential to examine the cell count **trend** to see

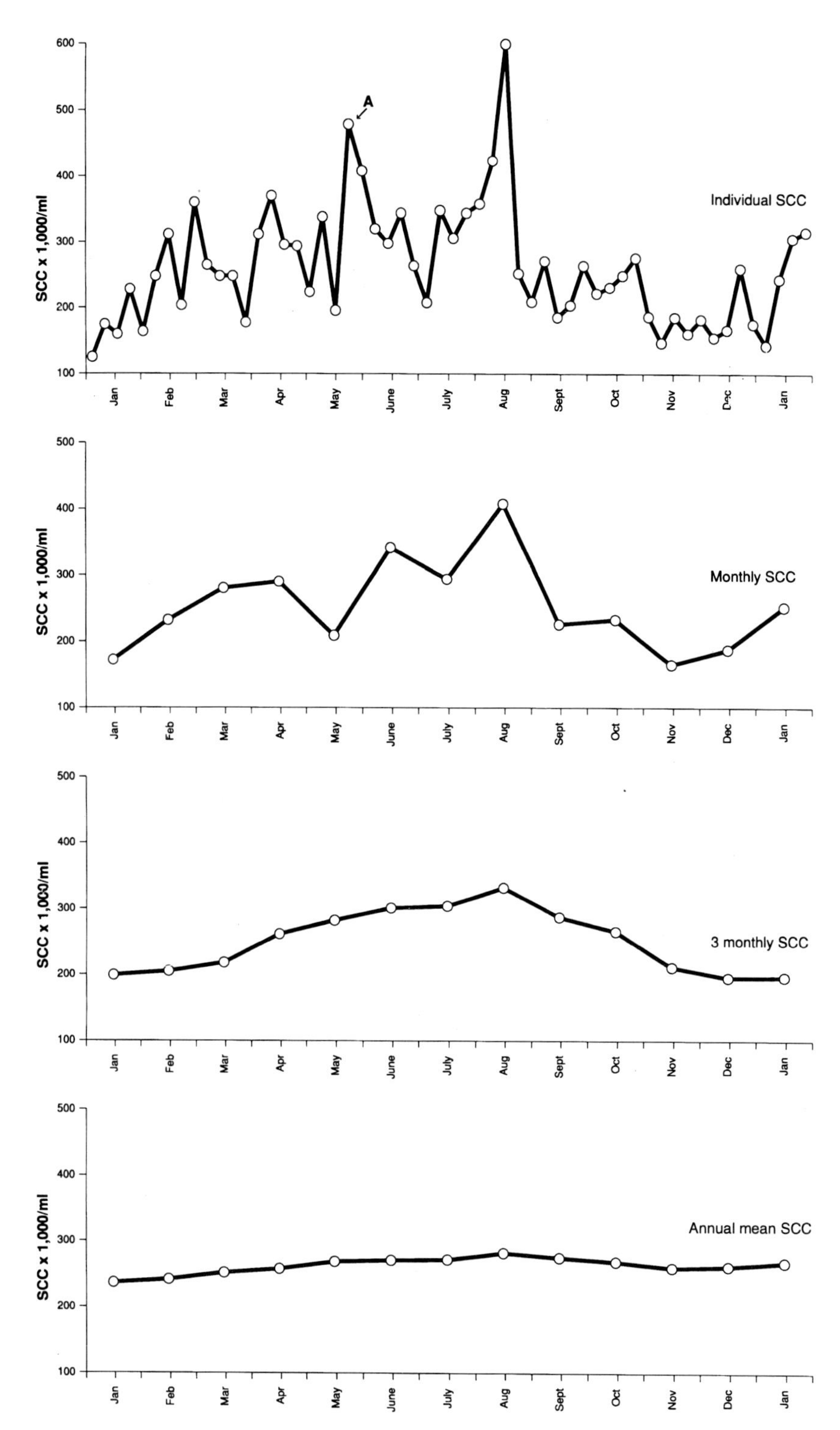

FIGURE 9.6 Individual, monthly, 3-monthly and annual rolling mean cell count (SCC) over a 13 month period.

what is happening in the herd.

High herd cell counts can only be reduced over a short period of time by ruthless culling of animals responsible for the increase. However, in the long term it is unlikely to solve the underlying mastitis problem. The dairy farmer who expects that he can reduce a cell count of 600,000 to 250,000 in a matter of a couple of months with minimal effort is likely to be disappointed. The speed of decline in most cases will depend on:

- the type of infection present.
- the proportion of the herd infected.
- how well control measures have been implemented.
- culling policy.
- financial situation of the farmer.
- the willingness to follow recommendations.

Very Low Herd Cell Counts

Can cell counts get too low? The simple answer to this is no!! At one time, it was felt that if the herd cell count was too low then cows would lose their ability to fight off infection that entered the udder and would therefore become more susceptible to environmental mastitis. This is not the case. It is the speed of movement of the white cells into the milk and not the number of white blood cells present before infection occurs, that determines whether or not bacteria will be eliminated (see page 26).

There is plenty of data available to show that herds with cell counts under 150,000 can have less mastitis than herds with higher counts. Data from 11 herds in a Somerset veterinary practice with cell counts under 70,000 had a low mastitis rate of between 7 and 21 cases of mastitis per 100 cows per year. This is well below the target figure of 30 cases per 100 cows per year (see page 142). However, it is important to remember that **some** low cell count herds may have more clinical cases of mastitis than a herd with a high cell count. This would be due to a high level of environmental mastitis. Table 9.2 shows some hypothetical examples of this.

Table 9.2 The incidence of contagious and environmental mastitis over a 12 month period in four herds with differing mastitis control and environmental management. (*29*)

Herds	*A*	*B*	*C*	*D*
Somatic cell count (× 1,000/ml)	125	125	600	600
Control of contagious mastitis	good	good	poor	poor
Environmental mastitis management	good	poor	good	poor
Contagious cases	7	6	25	24
Environmental cases	10	43	4	42
Total mastitis cases *	17	49	29	66

* All herds contained 100 cows.

While herd B has a low cell count it has a higher incidence of mastitis compared to herd C which has a higher cell count. The big difference between the two is the high level of environmental mastitis in herd B compared to herd C. The management of herd D leaves plenty of room for improvement. It has a high cell count which will result in reduced milk yield due to subclinical infection, and will be incurring financial penalties. The levels of environmental and contagious mastitis are both very high. Compare these figures to herd A with good mastitis management. As most low cell count herds are well managed, the risk of environmental infections is often kept to a minimum. It all comes down to attention to detail.

INDIVIDUAL COW SOMATIC CELL COUNTS (ICSCCs)

Individual cow cell counts are the best way to identify high cell count cows. Individual cell counts are calculated from a mixed sample from all four quarters. This may also be called a composite sample. Quarter cell counts refer to results from individual quarters. Samples for cell count testing need not be collected in a sterile manner. However, lumps of faecal matter may cause problems with electronic testing and foremilk should be discarded as it may have a higher count.

Individual cow cell counts are tested electronically and as it takes time from collection to the results getting back to the farmer – sometimes up to 14 days, when the farmer receives them they are historical data. This is to say that the results do not necessarily relate to current udder status.

In order to get the maximum benefit, cows should be sampled **regularly** so that averages can be studied rather than individual results only. A single high cell count indicates current infection status. However, subsequent tests may be low.

The danger of taking action on a single result has already been discussed earlier in this chapter. Many farmers with a high cell count have culled cows on the basis of a one-off screening of their herd only to find that the herd count has remained unchanged. Culling should **never** be considered on the basis of a single cell count.

Ideally samples should be measured every month. The average of the previous three months' results together with the lactational average should be considered before any action is taken. As the herd cell count rises, then so does the proportion of individual high cell count cows. In herds with a cell count over 700,000, it can be expected that over two-thirds of the cows will have cell counts over 500,000, compared to only 15% of cows in a herd with a cell count under 300,000.

Table 9.3 Guidelines for interpreting individual cow cell counts

Individual cow somatic cell count	*Udder health status*
Under 150,000	No infection likely
151,000–400,000	Probably infected
Over 400,000	Infection present

The big problem with individual cow results is that they do not identify which or how many of the **quarters** are infected, nor the level of any infection. This is shown in Table 9.4.

Table 9.4 The effect of quarter and individual cow cell counts (× 1,000/ml) in three cows

	Cow 3	*Cow 60*	*Cow 140*
Individual cell count	139	314	582
Interpretation	Not infected	Infection suspected	Infection present
Quarter cell count results			
LF (left fore)	20	600	425
RF	52	31	673
LH (left hind)	570	573	423
RH	33	51	807

From the composite result and interpretation guidelines shown in Table 9.3 we would expect Cow 3 to have no subclinical infection. The quarter results however show that there is significant infection present in the left hind quarter. Individual results and their interpretation from Cows 60 and 140 are correct.

USE OF INDIVIDUAL COW CELL COUNTS OR CMT RESULTS

Having accumulated and studied your cell count results over a period of 3–4 months, you need to know what action can be taken. There are a variety of options.

Culture

By sampling high cell count cows, the mastitis organisms present in the herd can be identified and specific control measures implemented. Samples must be collected carefully and submitted to the laboratory in the correct manner (see page 42).

As there is no method of knowing how many or which quarters are infected from an individual cow cell count result the use of individual quarter cell counts may be useful for cows with high results. However, this is costly and time-consuming. By far the easiest way to identify the quarters with high cell counts is to use the CMT test which can supplement or replace the use of individual cell counts. High count quarters can be checked on a couple of occasions to make sure that they are consistently high before collecting a sample for bacteriology.

Culling

This is a method of eliminating problem cows permanently, but it is costly due to the marked difference between the sale value of a cull cow and replacement costs. In addition, if the cull cow is replaced with a market cow, you may end up back where you started – with another high cell count animal. Few cows are sold through market because they give too much milk or have never had a case of mastitis!

Culling should **never** be based on cell count data in isolation. Factors such as the type of infection present should always be considered. For example, if a high cell count is due to *Streptococcus agalactiae*, then treatment will be successful in reducing the count. In this case, culling these cows would be a very expensive way of reducing the herd cell count. Of course, if infection is due to *Staphylococcus aureus*, then culling is a sound course of action.

Cows with persistently high cell counts for three or more consecutive tests must be considered for culling but it is **essential** to take other factors into account. These include:

- the number of cows in this category.
- the number of cases of mastitis that each animal has had.
- bacteriology results.
- milk yield.
- fertility status.
- general health.
- source of replacements.

Early Dry Cow Therapy

Early dry cow therapy should be considered if a high count cow is in late lactation. This will remove her milk from the bulk supply which will have an **immediate** effect in reducing the herd cell count. It will also remove the risk of spreading infection to clean cows.

Unfortunately dry cow therapy will not eliminate all infections. *Staphylococcus aureus* is the classic example (see page 152). If the infection is due to *Streptococcus agalactiae*, then dry cow therapy is very effective. Cows with an extended dry period may become overfat leading to calving problems and an increase in the number of metabolic problems during the next lactation.

Milking Order

High cell count cows act as a reservoir of infection. Milking these animals last should help to reduce the spread of infection. The big problem is the practicality of segregating and keeping these cows separate. At best it is difficult, if not impossible in many herds. Research work shows that this can be a relatively effective method of reducing disease transmission. If herds are grouped then the milking order is important (see page 91).

Treatment During Lactation

Treatment of subclinical *Staphylococcus aureus* mastitis during lactation is generally unsuccessful. This is because it tends to be chronic and well established. It is important to know which bacteria you are attempting to treat.

Cure rates are very low with *Staphylococcus aureus* infections (see page 152), often under 25% during lactation, and this combined with high treatment costs (intramammary antibiotics, discarded milk and extra labour) makes this line of action very expensive. The only time when treatment of *Staphylococcus aureus* during lactation may be considered is with an exceptional cow which gives 50% more milk than the herd average and if the farmer is prepared to accept that this form of treatment may well be unsuccessful.

If the high cell count is due to *Streptococcus agalactiae*, therapy may certainly be worthwhile. Unfortunately, chronic infections due to *Streptococcus uberis* are very difficult to treat.

Withholding Milk from the Bulk Supply

The herd cell count can be reduced by withholding milk from high cell count cows from the bulk supply. This will have an immediate effect in improving the situation and allows the farmer time to consider which line of action he wishes to take. In herds that are over quota limits, this temporary line of action can prove invaluable.

Some farmers will then feed this milk to calves. This is a controversial course of action – some suggest that mastitic milk may cause infection of the immature udder. Mastitis organisms may gain entry to the udder either by calves sucking the teats of other calves or they may be spread by flies.

Evaluating Treatment Efficacy

Cows with mastitis usually have increased cell counts. These counts will decline after successful treatment. In cases where there has been a bacteriological cure, i.e. all the bacteria have been eliminated from the udder, then cell counts would be expected to decline to below 200,000. If infection remains present and becomes subclinical then counts will remain elevated.

The benefits of dry cow therapy in eliminating subclinical infection can be demonstrated using individual cell counts. In an experiment 38 cows were sampled in the last two months before they were dried off using dry cow

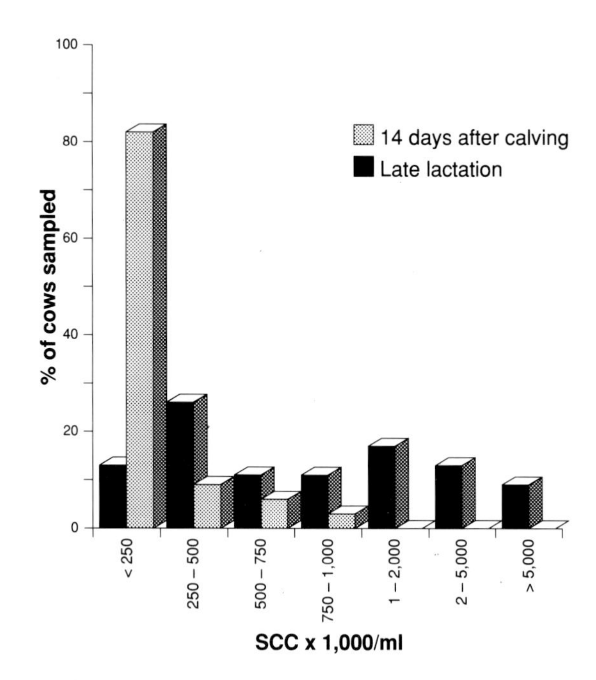

FIGURE 9.7 The use of individual cow cell counts (SCCs) to demonstrate the efficacy of dry cow therapy in reducing the level of subclinical infection at the end of lactation.

therapy. They were resampled 14 days after calving. The results are shown in Figure 9.7. It can be seen that over 60% of cows had a cell count over 500,000 before drying off compared to only 9% after calving, indicating that the dry cow therapy had removed the bulk of the subclinical infection at the end of lactation.

STUDY HERD

Table 9.5 shows some individual cell count data from a herd of 150 cows, with 130 animals in milk in the form used by NMR (National Milk Records) in the UK. The annual rolling herd cell count is 537,000. The bulk cell count on the day of sampling was 612,000.

General Considerations

Looking at the results, it can be seen that

Table 9.5 Individual cell count data (× 1,000 ml) for study herd of 150 cows

Cow no.	Lact no.	Days since calving	Total no. of tests this lactation	No. tests over 200,000	Tests results for: JUL	AUG	SEP	% Contribution to herd bulk tank SCC[1]
62	8	151	5	5	4670	2389	4388	6
77	5	253	9	5	413	581	3995	2
15	9	87	3	3	2049	2031	2879	7
82	8	132	4	4	1553	3605	2427	4
85	7	257	9	9	6796	1774	2391	3
21	8	27	1	1			2293	4
20	6	95	3	3	2024	1902	1805	4
73	8	149	5	5	3129	1484	1748	4
60	2	13	1	1			1609	3
69	6	476	15	14	1325	1458	1374	1
95	6	123	4	4	1315	753	1361	3
16	9	123	4	4	1655	1157	1360	2
94	6	208	7	7	620	792	1319	2
101	1	36	2	1		69	1221	3
28	7	236	8	7	6121	183	286	

[1] This will depend on both the individual cow cell count and her yield on that day.

15 cows (9% of the total number) were contributing 48% of all the somatic cells **on the day of recording**. If the milk from these animals had been withheld from the bulk supply on this day then the herd cell count would have dropped from 612,000 to just over 318,000. While this would have reduced the cell count, it would not have solved the underlying problem in the herd, responsible for the high count.

Now let's consider the individual cow results from four cows and the possible actions that could be taken.

Cow 62

This cow is in the middle of her 8th lactation. All five lactation tests have been over 200,000. The past three results have been very high and at the September recording, her milk made up 6% of the herd cell count. The milk from this cow must be withheld from the bulk supply.

Cow 15

This cow is 87 days calved in her 9th lactation. All three tests have been over 200,000. In September, her milk made up 7% of all the somatic cells in the herd. Like Cow 62, her milk must be withheld from the bulk supply.

Both cows 62 and 15 have had repeatedly high cell counts and their milk must be withheld from the bulk supply. Milk samples should be collected for bacteriology to identify or to confirm the cause of these high cell counts. This is essential in order that appropriate action can be taken.

Cow 101

This is a first lactation heifer that has been calved for only 36 days. The first cell count result in August of 69,000 indicated good udder health. The second result of 1,221,000 indicates infection, either subclinical or clinical. If the next set of results show a high cell count, then action will have to be taken.

Cow 28

This cow is in her 7th lactation and 236 days calved. The last two cell counts have been under 300,000. The result in July was very high at over 6,000,000. This is most likely to have been due to clinical mastitis on or before the day of sampling in July. If so, the two subsequent low results suggest that treatment has been successful.

These results show that the interpretation of individual cell counts requires **careful consideration**, and action can **only** be taken on a series of results, not on the basis of a one-off screening.

CHAPTER TEN

Total Bacterial Count (TBC)

The total bacterial count (TBC) of milk, is a measure of the bacteria grown from that milk over a fixed period of time. As the bacteria have to be grown to be counted, it is sometimes referred to as the total viable count or TVC. The results are given as the total number of bacteria per ml of milk. For simplification, these results are commonly reported back to the farmer in thousands, and so a count of 9 refers to a TBC of 9,000 per ml.

High TBCs affect the dairy farmer in two ways: directly in the form of financial penalties with the possibility of increased levels of mastitis, and indirectly through the production of a poor quality, short shelf life milk that is less acceptable to the consumer and manufacturer. Some producers believe that pasteurisation will not only kill all the bacteria present in milk, but will also put right any milk quality problems. This is not true.

In 1982 the MMB (Milk Marketing Board of England and Wales) started to measure TBC levels in bulk tank milk. A bonus payment scheme was introduced at the same time to promote the production of premium quality milk. Before the scheme, only 25% of producers had TBCs under 20,000. After it was launched there was a dramatic reduction with 61% of all producers having TBCs under 20,000 within four months.

The national TBC dropped from 22,000 in 1983 to 13,000 in 1994. Figure 10.1 shows that in 1994 over 85% of herds had a TBC under 20,000 compared to only 65% of herds in 1983. However in 1994, the average UK dairy producer still spent 2.4 months a year producing milk with a TBC level over 20,000. There is an escalating band of penalties imposed by dairy companies according to the TBC of the milk sold off the farm.

The 1992 EC Council Directive 92/46/EEC for dairy products required

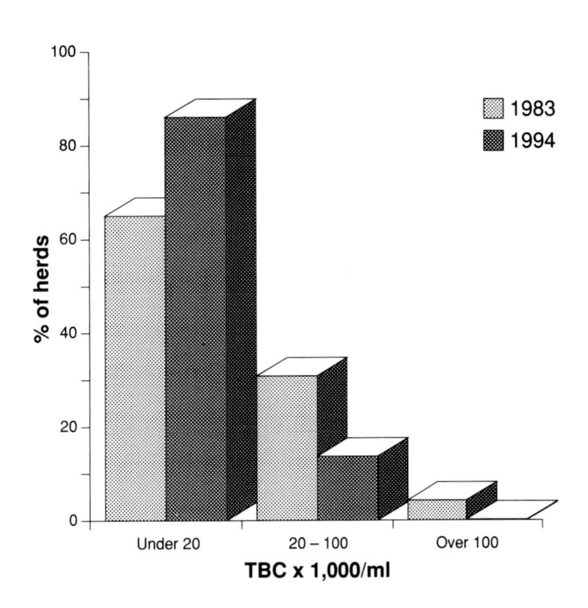

FIGURE 10.1 The distribution of TBC levels in the United Kingdom in 1983 compared to 1994. This shows the steady progress towards low TBCs.

TBC levels to be under 100,000 for liquid milk, and under 400,000 for milk for manufacturing, from January 1994. The level for manufacturing milk will drop to under 100,000 as and from 1 July 1997. It must be remembered that most dairy companies are only interested in purchasing 'quality' milk.

Overall, the position of the dairy industry in the UK appears very healthy: in 1994 99% of producers and 99.5% of milk met EC Directive standards. However, when you consider milk produced under good conditions will have TBC levels under 10,000, there is still plenty of room for improvement. There is no reason why any dairy farmer should not be able to produce milk with a TBC below 10,000 throughout the year. As increasing demands are placed on the dairy industry to produce milk of a higher quality, no doubt the threshold for TBCs will reduce further.

TBCs are measured regularly by the dairy companies. However, the benefits of their results to the farmer are limited as they do not identify the bacteria present nor the source of the organisms. Despite this, the TBC remains an accurate test to measure the number of bacteria in milk. With further testing, bulk tank samples can yield valuable information in relation to mastitis management.

Bacterial contamination of milk may occur in two ways: directly from the cow when mastitis organisms are shed into the milk, or indirectly from the environment or milking equipment.

THE FOUR SOURCES OF BACTERIA IN MILK

There are four main causes of high TBCs. These are:

- mastitis organisms
- environmental contaminants
- dirty milking equipment
- failure of refrigeration

Mastitis Organisms

Mastitis organisms should be suspected if the TBC fluctuates dramatically. Milk from a healthy quarter will have low numbers of bacteria present, usually under 1,000 per ml. When quarters become infected with clinical or subclinical mastitis the numbers of bacteria can increase substantially.

Streptococcus agalactiae and *Streptococcus uberis* are shed in extremely high numbers particularly from clinically and subclinically infected quarters, for example up to 100,000,000 per ml in clinically infected quarters. Large numbers of coliforms may also be shed with *E. coli* mastitis. In herds infected with these organisms it is easy to understand why TBC levels can fluctuate.

Take a 100 cow herd producing 1,500 litres of milk per day with an average TBC of 5,000. The addition of as little as 2 litres of mastitic milk from a clinical *Streptococcus uberis* cow (with a TBC of 10^8) can increase the TBC to 138,000. For this reason it is important to detect clinical infections early so that mastitic milk does not enter the bulk supply. Figure 10.2 shows the typical fluctuating effect of a *Streptococcus agalactiae* infection on the TBC in a herd of 50 cows. Other mastitis organisms, for example *Staphylococcus*, tend not to shed bacteria in large enough numbers to significantly affect the bulk tank TBC (see Table 4.3).

Unfortunately subclinical mastitis cannot be detected by the milker and so it is inevitable that some mastitis bacteria will enter the bulk supply. The best way to reduce this effect is through a mastitis control programme that will reduce the level of infection in the herd, and ultimately the number of mastitic organisms entering the milk.

Environmental Contamination

The main cause of environmental contamination of milk is inadequate teat preparation. The importance of good udder preparation has been referred to on page 79. It is essential that the milking unit is attached to clean dry teats. Milking dirty teats will not only contaminate the bulk milk but must also increase the likelihood of environmental mastitis.

The coliform count measures the number of coliform organisms in milk and gives an indication of the level of environmental contamination and the standard of pre-milking preparation. Coliforms are only one group of environmental organisms and there are many others such as *Streptococcus uberis*, *Streptococcus faecalis*, *Bacillus* species etc. The technique for carrying out a coliform count is described on page 45.

The target figure with good milking hygiene is to have a coliform count under 25 per ml, but levels under 50 per ml are acceptable. High levels of environmental bacteria will reduce the shelf life of milk and increase the risk of 'off' flavours and hence its acceptability to processors. Many cheese makers now carry out coliform counts on milk supplies daily to ensure that the milk will not affect their cheese making. There are many herds with excellent pre-milking preparation that regularly have coliform counts of zero.

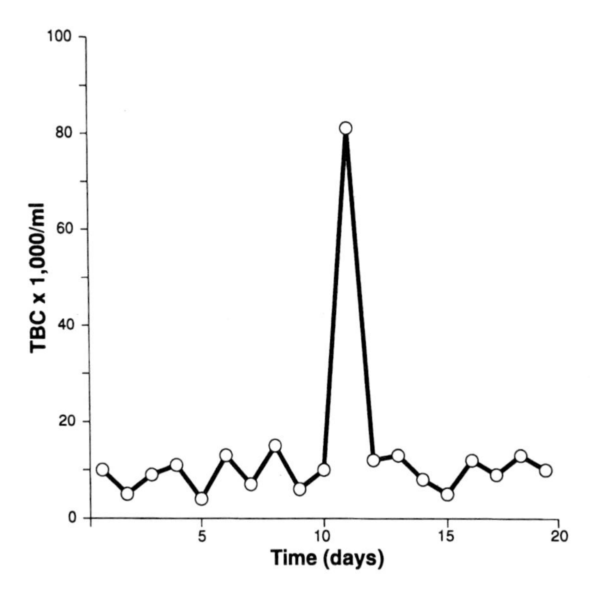

FIGURE 10.2 The typical effect that a single clinical case of Strep. agalactiae *mastitis can have on the bulk TBC in a herd of 50 cows.*

Table 10.1 shows a comparison of the TBC and coliform count before and after a change in the milking routine in a herd of 1,000 dairy cows in Arizona.

Table 10.1 The effect of different types of teat preparation on the TBC and coliform count/ml milk (*30*)

Test	*washing but not drying 3 months before change in routine*	*washing and drying 3 months after change in routine*
TBC	50,000	10,000
Coliform count	120	20

Initially teats were washed but not dried which resulted in high TBCs and coliform counts. Once the milking routine was modified and teats were washed **and** then dried before milking, the counts reduced significantly. Remember the coliform count will not measure all environmental organisms. It just gives an indication of whether the level of environmental contamination in milk is high or low.

The type of winter bedding used is also important: sawdust and wood shavings become rapidly contaminated with bacteria just 24 hours after bedding down – this is due to their large surface area and their ability to absorb moisture (see Figure 8.14).

In well-run dairy herds where there is plenty of clean, dry and well-bedded accommodation, teats should remain clean. In poorly managed herds where there is insufficient or dirty, damp accommodation, for example where there are uncomfortable cubicles resulting in cows lying outside, the condition of the teats entering the parlour will be a milker's nightmare.

Dirty Milking Equipment

Inadequately cleaned milking equipment can lead to raised TBCs. A laboratory assessment of plant cleaning can be made using the laboratory pasteurised count (LPC) or thermoduric count, where levels over 500/ml suggest a wash-up problem.

Milkers should look out for the following, which can cause contamination of the bulk milk tank:

- wash-up problems (boiler failure, build-up of milk soil in dead-end areas, blocked jetters or air injectors).
- dirty bulk tank: it should be inspected after every wash.

Wash-up routines are discussed on page 66.

Failure of Refrigeration

Milk should be cooled to 4°C or less **as soon as possible** after milking to limit the growth of bacteria. This helps to maintain milk life and is the maximum temperature at which the tanker driver is legally allowed to collect milk in the United Kingdom. When there is a refrigeration problem and milk is not kept cool, or cooled rapidly, bacterial multiplication will take place.

The importance of efficient refrigeration is becoming greater with a decreasing frequency of milk collection from some farms. In some countries, milk for liquid milk consumption is collected every second day, while milk for manufacturing is collected every third day, without any significant effect on TBC or milk quality **provided** the refrigeration is efficient.

The effect of multiplication will depend on the number and type of bacteria, together with the temperature of the milk. Warm milk is an excellent medium for bacterial growth. Some bacteria such as coliforms may double every twenty minutes under optimum conditions. This means that one coliform could multiply to more than seventeen million over an eight hour period. The increase in bacterial numbers in raw milk stored at different temperatures over a 12 hour period is shown in Figure 10.3.

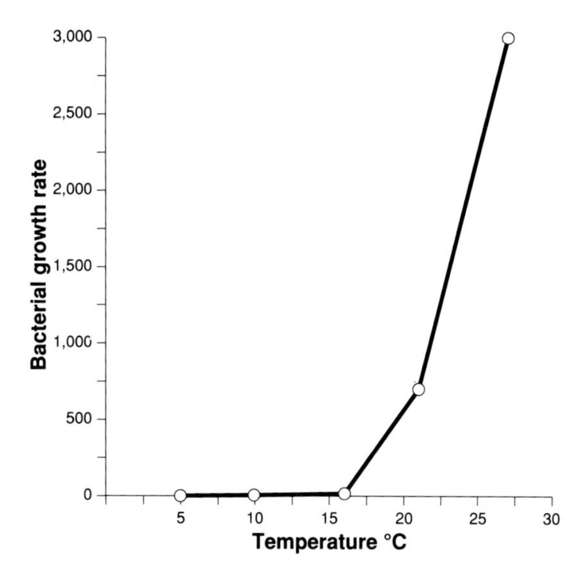

FIGURE 10.3 Effect of storage temperature on bacterial growth in raw milk over a 12 hour period. At 21°C there is a 700-fold increase in bacterial numbers over a 12 hour period. (31)

Plate coolers are commonly used to cool milk before it enters the bulk tank. They operate by using a heat exchange mechanism. Large volumes of cold running water (as much as seven times that of milk) flow in the opposite direction to milk with the heat from the milk passing through to warm the water, as shown in Figure 10.4. The resultant effect of the heat exchange mechanism should be to have milk leaving the cooler at a temperature as low as 6°C. Tube coolers (tubes surrounding the milk line through which cold water flows in the opposite direction to the milk) have the same effect.

The warm water from the coolers may be used to wash dirty teats before milking. Others divert this water to a drinking trough so that cows can have a warm drink after milking. Cooling milk before it enters the bulk tank saves energy as the tank has less work to do. It also helps to protect milk quality as the milk reaches 4°C more rapidly, thus reducing bacterial multiplication.

The shelf life of pasteurised milk is

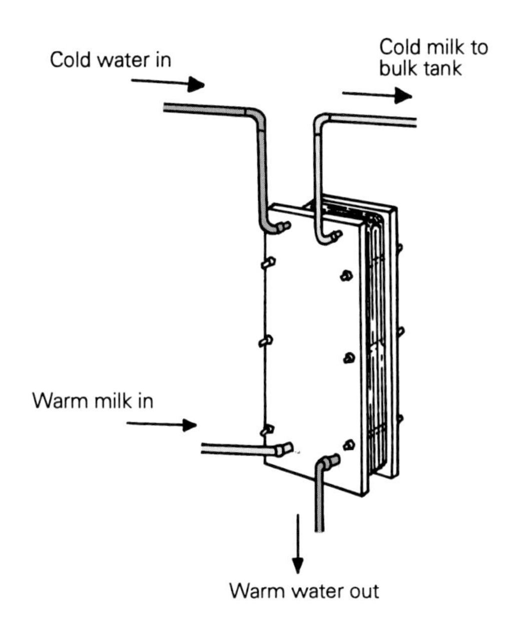

FIGURE 10.4 Plate cooler. Large volumes of cold water run through the plate cooler in the opposite direction to the milk, resulting in warm water but cool milk leaving the plate cooler.

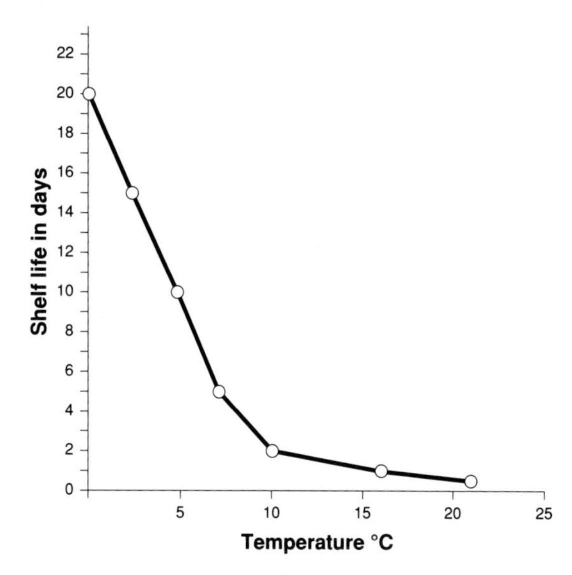

FIGURE 10.5 Shelf life of pasteurised milk in days according to the storage temperature. (31)

considerably affected by storage temperature as shown in Figure 10.5. This is of great importance to retailers, the milkman and of course the consumer. It can be seen that when pasteurised milk is stored at 16°C its shelf life is only one day compared to 10 days when stored at 5°C.

There is a variety of organisms that are not only capable of withstanding pasteurisation, but are also capable of growing under refrigerated conditions. These organisms are called **thermoduric psychotrophs** and grow best between 2 and 10°C. Some thermoduric psychotrophs such as *Bacillus cereus* are commonly found in the environment. If these enter the bulk supply, as may occur if teats are not properly prepared before milking, then they will multiply. Psychotrophs may adversely effect the shelf life and lead to off flavours in dairy products.

BULK TANK ANALYSIS

Taking a sample of milk from the bulk tank enables a variety of specific tests to be carried out to pinpoint possible management problems. It is an invaluable tool both for the vet and the farmer. Four tests can be carried out on bulk milk:

- TBC to measure the total number of bacteria present
- CC (coliform count) to check the level of environmental contamination and indicate the standard of pre-milking preparation
- LPC (laboratory pasteurised or thermoduric count) to assess the efficiency of the wash-up routine
- Identification of all bacteria present and hence their probable source, i.e. cow or environment

Bulk tank analysis is a useful way to identify herd and management factors associated with TBC problems. It may be possible to eliminate a dirty milking plant or environmental contamination as the cause of the high TBC. However, care is needed with interpretation. For example, if an organism has been isolated from a bulk sample, then this organism is present in the herd. However, if a suspected organism has not been identified, it does not mean that it is absent from the herd, but rather that it just has not been identified from that particular sample.

Action should never be taken on the results of one sample only as it may have been taken in unusual circumstances. For example, a relief milker may have been milking on the day of sampling, or a single cow with mastitis may not have been identified when mastitic milk entered the bulk tank.

It is **essential** that the sample milk is transported from the farm to the laboratory at no more than 4°C to minimise any bacterial growth. Milk should be fresh and not frozen, although collection of daily samples over a week, freezing and then processing the whole batch can help to eliminate laboratory variation associated with a series of samples. Freezing, however, will alter the number of organisms present, for example, there will be a reduction in the number of coliforms. If two bulk tanks are used, both should be sampled and tested individually.

Bulk tank samples should be collected in the following way:

- agitate the bulk tank for at least five minutes to ensure that the milk is well mixed
- scoop at least 30 ml of milk into a sterile sample pot using a sterile scoop or by wearing a clean disposable glove. This ensures that the remainder of the milk does not become contaminated during sampling.
- seal the container and label with the date and farm name, and tank identity (if samples are taken from more than one tank).
- store at 4°C from collection to lab: the sample can be kept in a fridge and then transported in an ice box to the laboratory.

Interpretation of Bulk Samples

The interpretation of bulk samples requires a knowledge of the various tests and how they relate to mastitis management. Frequently the problems identified may involve more than one area of mastitis. Table 10.2 shows the results of five bulk tank samples together with their interpretation. It must be remembered however, that bulk tank analysis is **only an aid** to identifying possible causes of high TBCs.

When investigating high TBC problems, a considerable amount of detective work may be needed in order to resolve the problem.

Result A

The TBC, CC and the LPC are all well below target. This is a good sample and indicates good pre-milking preparation and a clean milking machine.

Table 10.2 The results of five bulk tank samples compared with target levels

Test	*Target*	*A*	*B*	*C*	*D*	*E*
TBC (Total bacteria count) (×1,000/ml)	<10	3	21	29	150	12
CC (Coliform count)	<25	5	112	22	TNTC*	10
LPC (Laboratory pasteurised count)	<500	150	120	785	TNTC	95
Bacteria		*E. coli*	*E. coli* *Strep. uberis* *Strep. faecalis*	*E. coli* *Strep. faecalis* *Staph. aureus*	*E. coli* *Strep. faecalis* *Strep. uberis*	*E. coli* *Strep. uberis* *Strep. agalactiae*

* TNTC = too numerous to count

Result B

The TBC is higher than target as is the CC. The LPC is below target indicating that the wash-up routine is satisfactory. The high CC indicates that there is a lot of environmental contamination entering the milk. This is confirmed by the presence of *Streptococcus uberis*, *Streptococcus faecalis* and *E. coli*, all of which are environmental bacteria.

This is most likely to be caused by poor teat preparation, i.e. not cleaning dirty teats or washing but not drying teats. Another source of environmental contamination could be faeces sucked into the machine for example when a unit falls off a cow.

Result C

There is a high TBC and LPC. The CC is below target indicating good pre-milking preparation. The high LPC indicates there is a problem with the wash-up routine. Another contributing factor could be the presence of *Staphylococcus aureus* organisms entering the milk that may be adding to the effect of the elevated LPC.

As *Staphylococcus aureus* has been isolated, mastitis control measures against this organism should be checked.

Result D

The TBC, CC and the LPC are very high. This suggests that there are several problems which could be due either to poor pre-milking preparation, poor hygiene during milking or a dirty milking plant. However, it is more likely that there is a refrigeration problem or that the sample warmed up during transport to the laboratory, as all the test results are elevated. In this situation, check that the bulk tank is keeping the milk cool and collect and analyse another sample to see if the results are similar.

Result E

The TBC is slightly elevated but the CC and LPC are below target. *Streptococcus agalactiae* has been isolated in large numbers and this high TBC may have originated from an undetected clinical case or possibly subclinical infection. A further sample is required.

CHAPTER ELEVEN

Targets and Monitoring

Records are an important part in monitoring the incidence and severity of any disease. Mastitis is no exception. In fact, mastitis is one of the few diseases where a detailed analysis of the data can be used to help in the control of infection.

Many farmers rely on their cell count results to give an indication of their mastitis situation. Cell counts do give useful information but have limitations. High counts indicate the presence of subclinical mastitis, especially that due to *Staphylococcus aureus*, *Streptococcus dysgalactiae* and *Streptococcus agalactiae*. Unfortunately cell counts do not necessarily bear any relation to the clinical incidence of mastitis. Therefore, it is important to have and make use of accurate clinical records otherwise there is little benefit to be gained from keeping the data.

Mastitis records will enable the farmer to do the following:

- identify cows whose milk needs to be withheld from the bulk supply.
- identify problem cows who should be considered for culling.
- allow detailed monitoring of the herd mastitis performance to check that it is within acceptable limits and to see how the herd compares to others being monitored.
- gain valuable information that can point towards the possible cause of mastitis outbreaks and other problems.

RECORD KEEPING

The following should be recorded for each case of mastitis:

- cow number
- date
- quarter/s infected
- treatments given and number of tubes of antibiotic used
- bacteriology results (if available)

One case of mastitis is defined as one quarter infected once. A cow that calves down with mastitis in all four quarters, therefore counts as four cases of mastitis. If a treated quarter clears up but mastitis recurs seven or more days after the remission of clinical signs, then this is defined as a **new** case of mastitis. If mastitis recurs in the same quarter less than seven days after the remission of clinical signs, then this is defined as a **continuing** case of mastitis.

Mastitis cases can be recorded in a variety of ways (see Table 11.1). Ideally, subsequent cases of mastitis should be recorded adjacent to the first case so that problem cows are readily identified. In the first method of data layout shown in Table 11.1, it can easily be seen that cow 32 has had repeated cases of mastitis. Although exactly the same information is given using the second method, it is not immediately apparent that cow 32 is such a problem. Ideally, a cow's mastitis history should be kept on the same card for all lactations. Before drying off,

Cow	date	quarter	date	quarter	date	quarter	date	quarter
69	22.5.95	LH						
32	25.5.95	LH+RH	3.7.95	LH	15.8.95	LH+RF	23.10.95	LH
73	2.6.95	LF						
4	15.6.96	LF	17.9.95	RF				
17	12.8.95	RH+LH	21.9.95	RH+RF				
166	25.9.95	RH						
99	3.10.95	LF						
37	14.11.95	LH						
91	21.11.95	RH						

Table 11.1 Two forms of mastitis recording. The top system more readily identifies the problem cows as all information relating to the same cow is recorded on one line. In the chart to the right separate cases of mastitis in the same cow are not related back to each other as they are in the top chart.

Cow	date	quarter
69	22.5.95	LH
32	25.5.95	LH + RH
73	2.6.95	LF
32	3.7.95	LH
4	15.6.95	LF
17	12.8.95	RH + LH
32	15.8.95	LH + RF
4	17.9.95	RF
16	21.9.95	RH + RF
166	30.9.95	RH
99	3.10.95	LF
32	23.10.95	LH
37	14.11.95	LH
91	21.11.95	RH

records should be checked and cows with four or more cases of mastitis considered for culling from the herd.

Record analysis will allow the most appropriate control measures to be put into place. It is important that this data is analysed regularly. Every six months is ideal as it will help to identify possible problems and trends. Economic data can also be included to cost the benefits or losses from mastitis, together with TBC and cell count penalties.

WHAT DOES MASTITIS COST IN MY HERD?

It is possible to calculate what mastitis is costing in your herd over a year by gathering the data called for in Table 11.2 and using the formula provided.

First consider the total cost of mastitis per cow in the three herds shown in the table. Each herd has different mastitis management: herd A has excellent mastitis control. Herd B is an average herd where there is a cell count (SCC) penalty of 0.2 ppl (pence per litre) and a TBC penalty for one month's supply. Herd C is constantly penalised for a high cell count at 6 ppl (pence per litre) and is penalised for high TBCs for three months in twelve.

The results are quite surprising. Herd A has a small loss due to mastitis. Mastitis is costing over £76 per cow in herd B, which is a fairly average herd in England and Wales. The cost of mastitis in herd C is quite staggering at over £450 per cow. The bulk of this cost is due to cell count penalties equivalent to over £300 per cow. One wonders how much longer this farmer will be able to remain in production.

Next insert production data into

Table 11.2 An economic assessment of the annual cost associated with mastitis

		A	*B*	*C*	*Your herd*
HERD DATA					
A – Herd size		100	100	100	
B – Yield per cow (litres)		6,500	6,000	5,400	
C – No. of mastitis cases		25	45	80	
D – Cost per case of mastitis (£)		90	90	90	
E – Annual av cell count (× 1,000/ml)		175	320	519	
F – Milk price (ppl)		25	25	25	
G – Margin over concentrate (ppl)		18	16	15	
ECONOMICS					
H – Cost of clinical mastitis/year	C × D	2,250	4,050	7,200	
J – Loss due to subclinical mastitis (£)	$\frac{(E-200) \times 0.025 \times A \times B \times G}{10{,}000}$	Nil	2,880	6,278	
K – Cell count (SCC) penalty (£)	From the milk statement	Nil	1,200	32,400	
L – TBC penalty (£)	From the milk statement	Nil	125	337	
Total costs associated with mastitis (£)	H + J + K + L	2,250	8,255	46,215	
Cost (£) /cow/year	$\frac{(H + J + K + L)}{A}$	23	82	462	

* Milk production data refers to 1995 values, however, 1996 cell count penalties imposed by Milk Marque are used.

Table 11.2 for your herd and see how you compare.

It is unfortunate that because many of the costs of mastitis, especially those relating to subclinical infection, are 'unseen', they are not noticed by the farmer and so he sees little need to take action to improve his mastitis management.

MASTITIS TARGETS

Table 11.3 gives a range of figures which should be achievable within a herd, that is the targets to be aimed for and the level at which some action or interference should be taken.

Table 11.3. Target and interference levels for different mastitis and milk quality parameters

Targets	*Targets*	*Interference*
SCC	150,000	250,000
TBC	10,000	15,000
Mastitis rate (cases per 100 cows per year)	30	40
Percentage herd affected	20	25
Recurrence rate	10	20
Milking cow tubes per cow per year	1.4	2.5
Milking cow tubes per case	4.5	6.0
Percentage dry cow mastitis	1.0	2.5

Mastitis Rate

The mastitis rate is the number of cases of mastitis per 100 cows per annum. It is an invaluable measure of the mastitis incidence as it allows comparison between herds irrespective of size. The mastitis rate can be worked out using the formula below.

Mastitis rate =

$$\frac{\text{No. of cases of mastitis per year} \times 100}{\text{Total no. of cows in herd (milking and dry)}}$$

A high mastitis rate indicates a high number of mastitis cases in the herd but does not identify what type of infection is present, i.e. contagious or environmental.

Percentage of Herd Affected

The percentage of cows affected per year represents the proportion of the herd that have had one or more cases of mastitis over a 12 month period. This helps to give some indication of the type of mastitis present.

Chronic recurring mastitis, such as *Staphylococcus aureus*, could affect a small percentage of the herd, but there may still be a high mastitis rate. This may occur because the same cows keep getting repeat cases of mastitis in the same quarter. On the other hand an outbreak of environmental mastitis is likely to result in a larger percentage of the herd affected but relatively few repeat treatments in the same quarter.

The percentage of cows affected per year can be worked out using the following formula:

$$\text{Percentage cows affected} = \frac{\text{No. of cows that have had mastitis over a 12 month period} \times 100}{\text{Total no. of cows in the herd (milking and dry)}}$$

Recurrence Rate

The recurrence rate is the percentage of quarters requiring **one or more repeat** treatments over a 12 month period. A repeat treatment refers to one or more cases of mastitis recurring in the same quarter. The recurrence rate can be worked out using the following formula:

$$\text{Recurrence rate} = \frac{\text{No. of cases of mastitis requiring repeat treatment} \times 100}{\text{Total no. of mastitis cases}}$$

For example in Table 11.1 cows 32 and 17 have both had one or more repeat treatments. Cow 32 in the left hind (LH) and cow 17 in the right hind (RH). The total number of quarters needing one or more repeat treatments is therefore two – one for cow 32 and one for cow 17 (not four). If all four mastitis cases for the LH quarter of cow 32 were included, this single cow could skew the data. Chronic contagious mastitis, especially *Staphylococcus aureus* is mostly responsible for high recurrence rates.

Milking Cow Tube Usage

The number of milking cow tubes used in the herd will depend on the number of cases of mastitis and the number of tubes used to treat each case. A high number of tubes used per cow per year, e.g. over 3.0, could indicate one or more of the following:

- a high incidence of mastitis.
- infection that responds poorly to treatment.
- not all cases of mastitis have been recorded. As the total tube usage is not collected from farm data (it comes from your vet), this figure gives a useful indication of the accuracy of farm data.
- mastitis tubes used for reasons other than for treating mastitis!

Many veterinary practices now have computerised accounting systems that can give the number and type of intramammary tubes supplied over a given period of time. The number of milking cow tubes used per case of mastitis is worked out using the following formula:

$$\text{No. milking cow tubes per case} = \frac{\text{No. tubes used}}{\text{No. cases mastitis}}$$

$$\text{No. tubes per cow per year} = \frac{\text{No. tubes used}}{\text{Total no. of cows in herd}}$$

Seasonal Variation

It is useful to examine mastitis incidence by month of the year. A high number of cases in the winter months suggests a problem due to environmental mastitis, whilst all year round incidence points to a problem with contagious mastitis.

Age of Cow

Figure 11.1 shows the mastitis incidence broken down by lactation number in two herds. Herd Y shows infections spread evenly over all lactation numbers and this suggests a problem due to environmental mastitis. Herd Z, on the other hand, shows an increased incidence in older cows which suggests a problem with contagious mastitis.

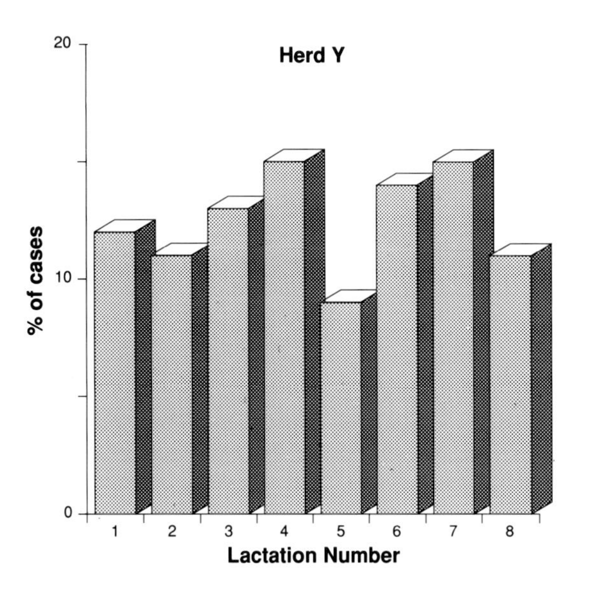

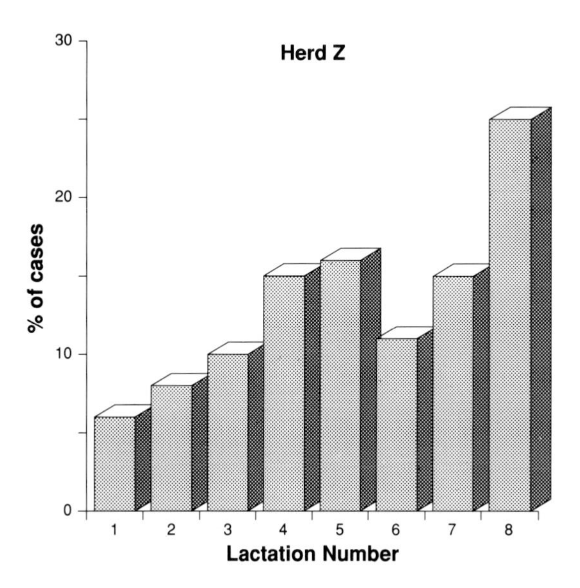

FIGURE 11.1 Clinical mastitis by lactation number for herds Y and Z. In herd Y the clinical incidence is equally divided between lactations, suggesting environmental mastitis. In herd Z, older cows have a higher incidence of mastitis, suggesting contagious mastitis.

Table 11.4. Mastitis and milk quality data for two herds together with acceptable target values.

	Herd A	*Herd B*	*Target*
SCC	632,000	171,000	150,000
TBC	14,000	5,000	10,000
Mastitis rate	60	57	30
Percent herd affected	16	31	20
Recurrence rate	22	5	10
Milking tubes per cow per year	4.2	2.3	1.4
Milking tubes per case	7.0	4.0	4.5
Seasonal difference	No	Yes	

Note that all the data shown in this table allows comparison between herds irrespective of herd size.

Herd A

The high cell count indicates a high level of subclinical infection in the herd, most likely due to *Staphylococcus aureus*, *Streptococcus dysgalactiae* or *Streptococcus agalactiae*. The TBC is above target and this may be due to increased numbers of mastitis bacteria entering the milk. This is more likely to occur with *Streptococcus agalactiae* than *Staphylococcus aureus* infection. The mastitis rate is high (60 cases per 100 cows per year) but only a small proportion of the herd (16%) is affected which indicates there must be several 'chronics'. Many cases of mastitis are recurring (22%) and there is a high tube usage per clinical case, perhaps due to poor response to treatment. There is no great difference in the mastitis incidence by month of the year. This would suggest a problem due to *Staphylococcus aureus* mastitis.

Herd B

This herd has a low cell count and a low TBC. This indicates a low level of subclinical infection. The mastitis rate again is high (57 cases per 100 cows per year) with a large percentage of the herd affected (31%). The recurrence rate of 5% however is very low. This suggests that a lot of cows are getting a single case of mastitis. The tube usage is a little above target but suggests no major problems with treatment. There is a marked seasonal incidence with a lot of cases of mastitis occurring during the winter months. This is what would be expected with an environmental mastitis problem.

HERD EXAMPLES

The mastitis data for herd A in Table 11.4 can be calculated for a 12-month period using the information below. This information will be available on most farms, but unfortunately is rarely analysed.

Herd size – all cows (milking and dry)	75
Number of cases of mastitis over 12-month period	45
Number of cows affected with mastitis over 12-month period	12
Number of quarters recurring (needing a 2nd or subsequent treatment)	10
Number of intramammary tubes purchased	315

$$\text{Mastitis rate} = \frac{45\ \text{(cases of mastitis)} \times 100}{75\ \text{(cows in herd)}} = 60\ \text{cases per 100/year}$$

$$\text{Percentage of herd affected} = \frac{12\ \text{(No. of cows affected)} \times 100}{75\ \text{(cows in herd)}} = 16\%$$

$$\text{Recurrence rate} = \frac{10\ \text{(No. of quarters requiring a 2nd or subsequent treatment)} \times 100}{45\ \text{(no. of cases of mastitis)}} = 22\%$$

$$\text{No. of milking cow tubes used per cow per year} = \frac{315\ \text{(milking cow tubes used)}}{75\ \text{(herd size)}} = 4.2$$

$$\text{No. of tubes used per clinical case} = \frac{315\ \text{(milking cow tubes used)}}{45\ \text{(No. of cases of mastitis)}} = 7.0$$

CHAPTER TWELVE

Mastitis Treatments and Dry Cow Therapy

While much of this book focuses on the prevention and control of mastitis, the text would not be complete without some reference to treatment. The main objective of treatment is to reduce or eliminate infection from the udder. Treatment can be administered at two different stages in the cow's lactation cycle and in three different ways:

- **lactating cow therapy** is administered to cows while they are in milk.
- **dry cow therapy** is aimed at removing any infection present in the udder at the end of lactation (i.e. to prevent carry-over to the next lactation) and reducing the number of new infections contracted during the dry period.

Treatment can be administered by different routes:

- **intramammary treatment** is infused into the udder through the teat canal.
- **parenteral treatment** is given by injection.
- **oral therapy** (drenching) is given in liquid form by mouth.

Irrespective of the cause of mastitis, there are several reasons why some form of treatment (not necessarily antibiotic) should be instigated. These are:

- to restore the productivity of the cow, thereby allowing her milk to be sold as soon as possible.
- to prevent the mastitis from getting any worse.
- to avoid long-term and possibly irreversible udder damage, which would have a deleterious effect on yield and may affect milk quality (i.e. cell count and TBCs).
- to prevent the spread of infection to other animals.
- to improve the overall health and hence the welfare of the cow.

TREATMENT DURING LACTATION

It is the milker who will first recognise a case of mastitis and who must make the many decisions relating to treatment. This chapter is therefore written with this person very much in mind. Foremilking and other procedures to assist in the prompt recognition of mastitis are described in Chapter 6. Ideally, once identified, the mastitic cow should be separated from the herd and treatment administered.

Separation from the Herd

This may consist of physically removing the cow to a mastitic or 'hospital' group of cows, which is then milked last. However, in smaller herds it is more common to milk affected cows through a separate cluster and into a dump bucket (Plate 12.1).

The reasons for the separation of the mastitic cow are:

PLATE 12.1 A dump bucket with separate cluster.

- an infected cow (and especially one infected with *Staphylococcus aureus*) may, through contaminated teat liners, transmit infection to the next six to eight cows milked. Milking her last, or through a separate cluster, avoids this. However, it is vital that the mastitic cluster is disinfected, e.g. by immersing it in hypochlorite solution, before milking the next mastitic cow, otherwise infection can still be spread.
  This is often overlooked and is particularly important if, as is often the case, the same cluster is also used to milk freshly calved cows, whose milk is being discarded because of colostrum or dry cow antibiotic.
- using a separate bucket and cluster, there is a lesser (or zero) risk of contaminating bulk milk with antibiotic from treated quarters. Some parlours have a dump line, through which colostrum and mastitic milk pass into a separate collection vessel. However, they still may have no separate cluster. This is very dangerous – you would need to avoid very few cases of mastitis to pay for a spare cluster!

Administration of Intramammary Antibiotics

This should be done as carefully and as cleanly as possible. Rough handling can lead to teat canal damage, which in turn predisposes to mastitis (see Chapter Three). Administration of antibiotic through a contaminated teat end might introduce a yeast infection which is particularly difficult to cure. The following procedure is suggested:

1. ensure that both the milker's hands and the affected teat are clean. Dry wipe both with a clean paper towel and wash, if necessary.
2. swab the end of the teat with meths or alcohol, until it is clean, i.e. until the swab can be rubbed across the teat end without becoming soiled (Plate 12.2).

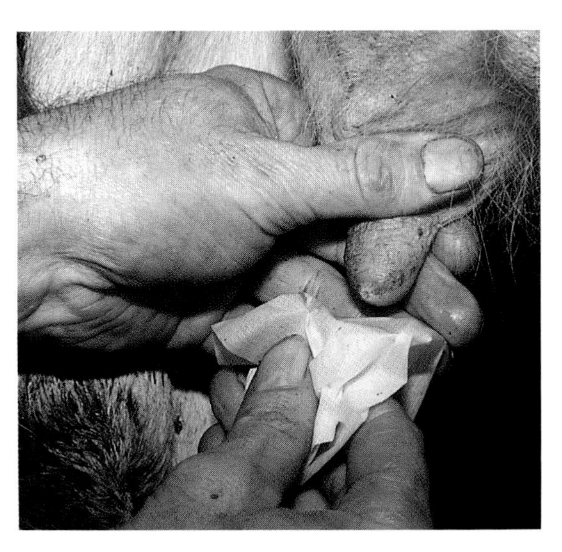

PLATE 12.2 Swab the teat end until it is clean.

3. remove the cap of the antibiotic tube and, without touching its tip with your hand, gently insert it into the teat

canal (Plate 12.3). It is **not necessary** to insert the nozzle to its full depth: in fact to do so could dilate the teat canal excessively, thereby cracking its protective lining (see page 17) and predisposing the cow to mastitis. Partial insertion is also recommended (particularly for dry cow therapy) because it enables some antibiotic to be left in the canal itself.

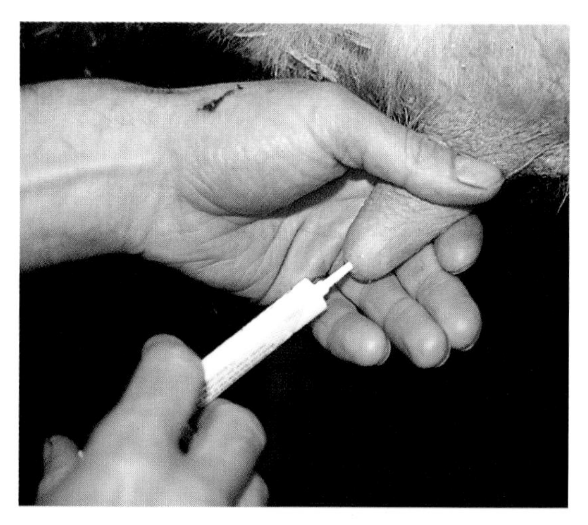

PLATE 12.3 Inserting an antibiotic syringe.

Some manufacturers are now producing antibiotic tubes with very small nozzles to achieve partial penetration and reduce teat-end damage (see Figure 12.1). However, if the cow is nervous or difficult to handle, full penetration may be unavoidable.

4. dip all teats after administration. This is important for both lactating and dry cow therapy. Tubing the cow, however carefully, dilates the teat canal and hence the extra protection of a dip is very valuable. In addition, it is probable that even if only one quarter has mastitis, during the milking process infection may spread to the teat orifice of the other three quarters. This infection can be removed by thorough teat disinfection.
5. carefully mark the cow to show that she has been treated with antibiotic. This reduces the risk of antibiotic contamination of bulk milk. A variety of marker sprays, leg tapes and tail bands are used.
6. record the treatment in the Medicines Book (a legal requirement in the UK) and elsewhere as necessary (see Chapter 11).

Is Antibiotic Treatment Worthwhile?

There is a body of opinion which considers that antibiotic treatment of some types of mastitis during lactation is simply not worthwhile. The reasons given for this are:

- the response of *Staphylococcus aureus* infections to treatment is very disappointing (see Table 4.2 and page 31) – although the clots and other clinical signs may disappear, there may be only a 20–35% bacteriological success rate.

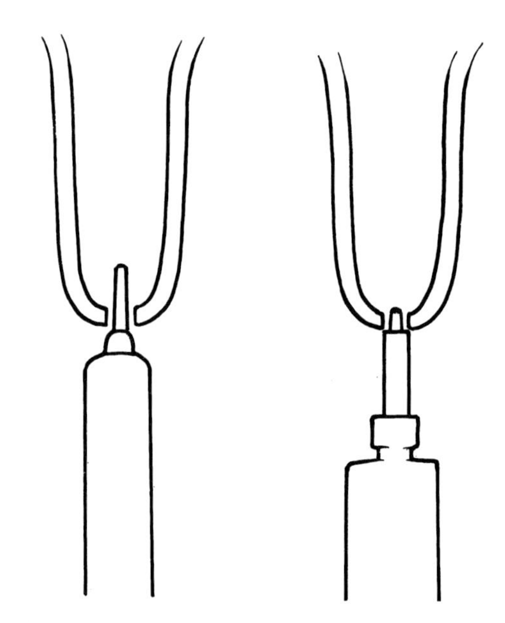

FIGURE 12.1 Long and short nozzle tubes: the short nozzle (right) is preferable, as it will prevent unnecessary damage of the teat canal.

- many cases of mastitis undergo self-cure – the infection is naturally eliminated by the cow without treatment. This is particularly likely with coliform infections, where the response by the cow can be so dramatic that in some cases all bacteria may have been eliminated within 4–6 hours (see page 24). However, even coliforms occasionally establish themselves as chronic persistent udder infections.
- the cost of the discarded antibiotic milk, plus the risk of antibiotic contamination of the bulk milk, are both so high that it renders treatment uneconomic.

Many excellent papers have been written on this subject, some in favour of treatment, others against. It is the opinion of the authors that, on balance, treatment is worthwhile. Response of *Streptococcus agalactiae* and *Streptococcus dysgalactiae* to treatment is generally good, although it is accepted that for *Staphylococcus aureus* infections complete bacteriological response can be very poor. Table 12.1 shows that for both *Staphylococcus aureus* and coliforms the spontaneous cure rate may be only slightly lower than that following antibiotic therapy. The highest rate of antibiotic cure for *Staphylococcus aureus* relates to an initial infection. More persistent and chronic infections achieve a lower success rate.

Table 12.1 Spontaneous versus antibiotic cure rates for mastitis. (*32*)

	Spontaneous cures	*Antibiotic cures*
Staph. aureus	20%	20–35%
Strep. agalactiae and *dysgalactiae*	19%	36–95%
Coliforms	70%	71–90%

We have seen throughout the preceding sections that mastitis control is very much a 'numbers game', which involves reducing the level of the bacterial challenge at the teat end, rather than totally eliminating it. Antibiotic treatment is one further way of reducing the challenge, this time within the udder. While it may not eliminate infection totally, it may well reduce the threat to manageable proportions. In addition, if antibiotic therapy saves one death from peracute mastitis, it offsets the cost of a large number of unnecessary treatments.

CHOICE OF ANTIBIOTIC FOR TREATMENTS

This is a huge subject and in itself consists of sufficient material to fill a single book. This section gives simple guidelines only. It does not lay down specific rules for treatment, but rather points to the complexity of the subject and gives examples of a few of the factors involved.

As is the case with the purchase of a car, there are numerous manufacturers, each with their own range of products and each with its own unique claims of effectiveness. Also, like cars, there are many products on the market between which there is little to choose in terms of value for money.

The following criteria should be considered when making a choice of antibiotic for treatment:

- antibiotic sensitivity of the bacteria involved.
- ability to penetrate the udder.
- ability to persist in the udder at a concentration sufficient to kill bacteria following single or multiple infusions.
- effectiveness in the presence of milk.
- whether it is bacteriocidal (killing) or bacteriostatic (arresting growth) (see page 151).
- lipid solubility, plasma protein binding properties and pH level in solution.
- withdrawal periods.
- cost.

There is an excellent article by MacKellar (*33*) which gives much more detailed information.

Antibiotic Sensitivity and Udder Penetration

Penicillins

As a general rule, penicillins are effective against gram-positive bacteria (staphs and streps, see page 43), but not against gram-negatives (coliforms etc.). Most penicillins penetrate the udder well. Examples include:

- penicillin G
- penethamate
- cloxacillin
- nafcillin

Unfortunately many (approximately 70%) mastitic strains of *Staphylococcus aureus* are now penicillin-resistant, because they have adapted to produce the enzyme beta-lactamase. However, cloxacillin and nafcillin are effective in the presence of beta-lactamase and, to date, **no staphylococci** have been found which are resistant to these drugs. Cloxacillin and nafcillin are therefore suitable preparations for dry cow therapy (a large percentage of infections present at drying off are caused by staphylococci (see Table 12.6). However, they are not effective against coliforms, which are gram-negative.

Some synthetic penicillins have been modified so that they are effective against coliforms, namely:

- ampicillin
- amoxycillin

However, these two drugs are still not effective against beta-lactamase staphylococci. Clavulanic acid is an irreversible inhibitor of beta-lactamase and by combining clavulanic acid with amoxycillin, one manufacturer has produced a product which should be effective against the vast majority of mastitic bacteria. Other combination products which should, theoretically, achieve good udder penetration and be effective against all organisms are a mixture of cloxacillin (kills gram-positives and beta-lactamase staphylococci) plus amoxycillin (kills gram-positive and gram-negatives), or cloxacillin plus ampicillin.

Erythromycin

Erythromycin penetrates the udder well, has good action against gram-positive bacteria, including beta-lactamase staphylococci, but it is not effective against gram-negatives. It also has a short milk discarding period post-treatment.

Aminoglycosides

The aminoglycoside group of antibiotics, namely:

- streptomycin
- neomycin
- framycetin

are active against coliforms, and are effective against beta-lactamase staphylococci. They have poor penetration of the udder tissue. One of their strengths is that they are relatively inexpensive. As penicillins achieve good penetration of the udder, products containing penicillin and streptomycin are often used in combination.

Cephalosporins

Cephalosporins are active against gram-negative and gram-positive bacteria, including beta-lactamase staphylococci, although penetration of the udder is not as good as with the penicillins. 'Second generation' cephalosporins, for example cefuroxine, have improved activity against gram-negatives, whilst 'third generation' products, for example cephoperazone, have the added advantage of being effective against *Pseudomonas*.

Tetracyclines

Tetracyclines are broad-spectrum, that is they are effective against gram-negative and gram-positive bacteria, with some activity against beta-lactamase staphylococci. However, penetration of the udder tissue is limited (although this may be overcome by using very high dosages) and resistance may occur with coliforms.

Response of Coliforms to Antibiotics

Coliforms have a variable sensitivity to antibiotics and standard texts (*34*) show how wide this can be. For example, in two reports the range of sensitivity to tetracycline varied from 23%–68%. For ampicillin the range was 35%–64%. As a herd outbreak of *E. coli* mastitis (the most common of the coliform types) always involves a range of **different** strains of *E. coli*, precise guidelines regarding the most effective antibiotic to use on the basis of the sensitivity of a single bacterial isolate cannot be given. Gentamycin is extremely effective against both *E. coli* and *Klebsiella*, but the cost of treatment is usually prohibitive, with a very long withdrawal period.

General Considerations

Whichever preparation is decided upon for routine use (and this must be a joint decision between the farmer and his vet), the following factors are important:

- due to the increasing incidence of coliform mastitis (see Table 4.1) lactation treatments should always involve a broad-spectrum antibiotic, e.g. one which is effective against gram-positive (i.e. staphs and streps), gram-negative (i.e. coliforms) and beta-lactamase organisms.
- preparations used in dry cow therapy should be aimed primarily at staphylococci and streptococci.

Effectiveness in Milk

Although the antibiotic sensitivity plate test (page 43) is commonly used to assess the response of bacteria to antibiotics, many antibiotics are much less effective in the presence of milk than this test suggests. For example, the ratio for oxytetracycline is 4:1. This means that oxytetracycline is four times less effective in the presence of milk than on the plate test. Other examples include streptomycin (5:1), erythromycin (7:1) and trimethoprim/sulphadiazine (500:1). However, these figures were obtained in experiments using whole milk and so the ratios may not apply to mastitic milk which has a higher pH than uninfected milk.

Bacteriocidal and Bacteriostatic Antibiotics

Antibiotics vary in the way they act. Some, for example the penicillins, specifically kill bacteria (**bacteriocidal**). However others, for example the tetracyclines, simply prevent bacterial growth and multiplication (**bacteriostatic**) and rely on the cow's own defence mechanisms to overcome the infection. If the cow is very sick, the activity of her defence mechanisms may be compromised and in such cases the use of bacteriocidal antibiotics may be preferable. However, the counter argument to this is that bacteriocidal antibiotics may lead to sudden bacterial death and release of endotoxins (especially with coliform infection, see page 25), making the condition worse.

Acidity and Lipid Solubility

Antibiotics are either acidic or basic, depending on their pH when in solution. Because the pH of milk (6.7) is lower than the pH of blood (7.4), drugs such as erythromycin and trimethoprim which are naturally more alkaline will be drawn into the mammary gland and therefore penetrate the udder well. However, when the udder becomes inflamed, as in severe mastitis, the pH of milk increases towards the pH of blood and this factor becomes less important. The lipid solubility and the degree to which antibiotics bind to proteins in blood will also affect their ability to penetrate the udder, particularly following intravenous or intramuscular injection.

Withdrawal Period

Post-treatment milk and meat withholding periods are stated on the product and must always be observed. Although the majority of the antibiotic remains in the treated quarter, some will diffuse into the bloodstream, pass around the body and be deposited back into the untreated quarters. This is because there is a very high blood flow through the udder (400–500 litres of blood for each litre of milk produced). When the affected quarter is inflamed flowrates may be even higher. Milk must therefore be discarded from all four quarters, even if only one quarter is being treated.

The withdrawal period given on the tube relates to the use of that tube as stated in the instructions. If the herdsman decides to use an increased frequency of tubing, or administers two tubes at the first treatment, or injects the cow with antibiotic in addition to tubing her, then this could affect the required withdrawal period. However, it is a difficult area and no precise guidelines can be given. If in doubt, ask your vet.

Ability of Antibiotics to Persist in the Udder

The ability of an antibiotic to persist in the udder at bacterial killing concentrations depends partly on the chemical nature of the antibiotic and partly on its formulation. For example, products with a long persistency, as would be required for dry cow therapy, are formulated in slow release oils or waxes, or manufactured with a much smaller particle size. Conversely, aqueous preparations are generally shorter-acting, with a low persistency, but short milk withholding period.

Antibiotic by Injection Concurrent with Intramammary Tubing

Despite the problems of withdrawal periods noted above, antibiotics by injection are commonly administered at the same time as intramammary tubes. This is particularly common if a cow is very ill, or has an elevated temperature. Some research has shown that the treatment of mastitis is equally effective whether antibiotic is administered as an intramammary tube or by injection.

A recent trial demonstrated the advantages of injecting penicillin at the same time as using amoxycillin intramammary tubes (Table 12.2). The trial cows were infected with *Staphylococcus aureus.*

Table 12.2 Advantages of injecting penicillin concurrently with amoxycillin intramammary tubes (*35*)

	3d. Amoxycillin intramammary	*3d. Amoxycillin intramammary + penicillin i/m*
Number of quarters	40	35
% cured	25	51

This treatment would not be effective against beta-lactamase-producing staphylococci.

Aggressive antibiotic therapy, for example five to seven days of concurrent intramammary tubes and injectable amoxycillin and clavulanic acid or tylosin, has also been suggested for the treatment of chronically infected cows with a high cell count. Others have suggested five days of 'milking cow' intramammary treatment, followed by drying off and dry cow treatment. Neither of these treatments are highly effective, however, and culling is still the only certain way of removing chronic carriers.

Resistance of *Staphylococcus aureus* to Treatment

Staphylococcus aureus (also known as coagulase positive staphylococci) gives a notoriously poor response to treatment. This was demonstrated in Table 4.2 and again in Table 12.1. Even with dry cow therapy, response is poor (Table 12.6). This is particularly the case in older cows, where the infection has been present in the udder for some time (Table 12.7).

There are several reasons for the disappointing response of staphylococci to treatment. These include:

- *Staphylococcus aureus* forms abscesses within the udder. A typical example is shown in Plate 4.2. These abscesses are often surrounded by a thick fibrous capsule. This prevents the antibiotic from reaching the bacteria, or insufficient antibiotic concentration is achieved within the abscess to kill bacteria effectively.
- Some strains of *Staphylococcus aureus* can live within cells such as macrophages (see page 21). Most antibiotics are only able to circulate in the body fluids surrounding cells and not able to penetrate the cell itself. Those staphylococci which live inside the cells are hence protected from the majority of antibiotics. (A few of the newer antibiotics can penetrate cells, but these are not yet available as mastitis preparations). *Staphylococcus aureus* can be released from the cells at a later date to cause further bouts of mastitis.
- Many strains of *Staphylococcus aureus* produce beta-lactamase, making them resistant to certain types of penicillin. Even when effective antibiotics are used however, response to treatment is still very poor.
- Some strains of *Staphylococcus aureus* can persist in a state of bacterial dormancy, that is they completely cease multiplying. In this state they are not killed by antibiotics, although they can become active at a later date.
- L forms of *Staphylococcus aureus* may occur. These are bacteria which do not have a proper cell wall and therefore most antibiotics will not kill them. This includes cloxacillin and other antibiotics which are effective against beta-lactamase strains. (The antibiotic novobiocin is effective.) However, there is some doubt whether L forms of *Staphylococcus aureus* can ever be produced under the conditions present in the udder.

One of the difficulties with assessing response to treatment for *Staphylococcus aureus* is that following a course of antibiotic therapy, many quarters initially appear to have responded and no bacteria are isolated from a milk sample.

However, this is simply because no *Staphylococcus aureus* are present in that particular sample. If the same cow is sampled at a later date, bacteria may have been released from an abscess or revived from their dormant state and it can be seen that the cow is still infected, i.e. treatment has not been effective.

This is clearly demonstrated in Table 12.3.

Table 12.3 The results obtained in trials assessing the response rate of *Staphylococcus aureus* to treatment with antibiotics (*36*)

No. of days after treatment	*% cows with Staphylococcus aureus*	*% response to treatment*
16	43	57
30	56	44
60	62	38

When cows infected with *Staphylococcus aureus* were sampled 16 days after treatment, it was found that bacteria were still present in only 43% of treated quarters, so the success rate was 57%. However, if sampled again at 30 days, bacteria were isolated from 56% of the treated cows, and this increased to 62% of cows (only a 38% response) if sampled at 60 days post treatment.

These results were obtained in a trial using combined intramammary and injectable antibiotics. If only intramammary treatment was used, then the response at 60 days was even lower (27%).

Blitz Therapy Against Streptococcus Agalactiae

The response of *Streptococcus agalactiae* to treatment is totally different to that of *Staphylococcus aureus*. *Streptococcus agalactiae* is very sensitive to almost any antibiotic and response to treatment, even during lactation, is generally very good (Tables 12.1 and 4.2). This allows a system known as 'blitz therapy' to be used in the elimination of *Streptococcus agalactiae* from milking herds.

Blitz therapy involves the use of intramammary antibiotics infused into all four quarters of the udder. Two forms are used: total – where the entire milking herd is treated – and partial – where only selected animals are treated.

In order for this to be successful, *Streptococcus agalactiae* must be the organism responsible for the mastitis problem. The herd owner must organise the milking routine so as to minimise any possible spread of infection throughout the herd. Post-dipping is **essential**. All dry cows must receive dry cow therapy and the approach must be economically viable.

The future source of replacements must be considered, as they may be a possible cause of re-infection. For this reason many recommend that newly purchased cows are treated with intramammary antibiotics before they join the milking herd.

Blitz therapy is not always successful. There are many reasons for this, for example:

- infected cows may be introduced to the herd.
- the milkers become careless and allow any residual infection to be transferred from cow to cow.
- dry cows may not have received dry cow therapy and so re-introduce infection to the milking herd.
- when using selective blitz therapy, not all infected cows may be identified for treatment and so infection may remain in the herd.

SUPPORTIVE THERAPY

In addition to antibiotics, a wide range of other treatments have been suggested for different types of mastitis. This section is not intended to turn farmers into vets, but rather to give an insight into some of the options available and into the complexity of the subject.

Fluid Therapy

Toxins, particularly those produced by coliforms, can cause a state of shock, and affect many body organs. Blood vessels dilate and as a result blood pressure starts to fall. Falling blood pressure leads to poor circulation and consequently poor tissue perfusion with blood. The animal appears dehydrated, as its body fluids are in the tissues and not in its circulation. Dehydration may frequently reach 7–10% of body weight. This means that for an average 600 kg cow, 40–60 litres of fluid need to be replaced to restore the circulation to normality. Dehydration further increases the feeling of malaise and general ill-health of the cow. Administration of fluids, especially in animals which will not drink, can be of considerable benefit. In fact some people are of the opinion that fluid therapy alone, without antibiotics, is sufficient treatment.

Fluids may be administered in a variety of ways:

Intravenous administration

In the field this is often carried out using a small garden pump (Plate 12.4). As the rate of administration is slow, this can be a time-consuming and therefore expensive exercise. Firm fixation of the intravenous catheter is required, and as such should only be undertaken by vets. An effective intravenous solution can be prepared by mixing a proprietary packet of electrolyte powder with warm tap water. This should help to restore normal metabolic activity.

Some people recommend the

PLATE 12.4 A garden pump is often used for administering intravenous fluids.

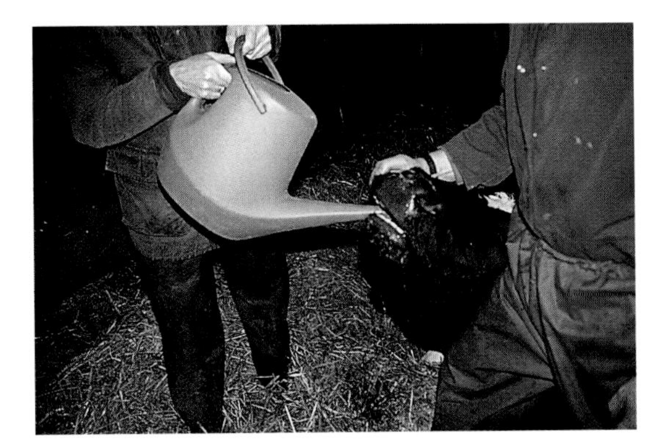

PLATE 12.5 Oral fluids are easily administered using a watering can.

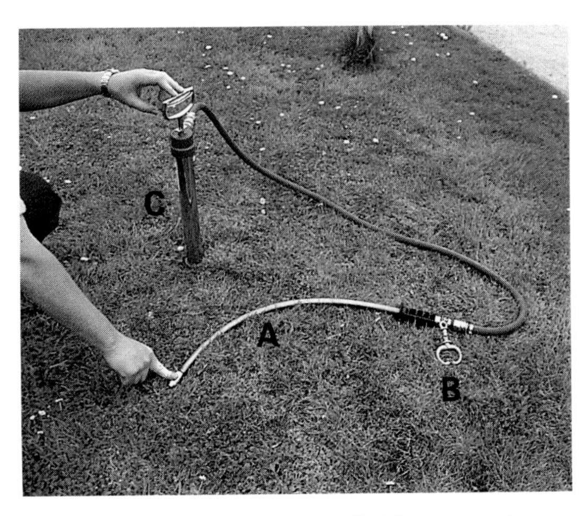

PLATE 12.6 Oral pump fluid apparatus: (A) flexible metal tube to insert through mouth and into oesophagus, (B) nose clips hold the tube in position, (C) pump which is placed into the bucket of oral fluids.

intravenous administration of 2.0 litres hypertonic (concentrated) salt solution (70 g per litre NaCl). This stimulates extreme thirst in the cow and encourages her to drink. However, if the technique is used in recumbent cows, it is obviously **vital** that water is fully accessible. Extreme care is needed during the infusion to monitor for shock.

Oral administration

Provided that the cow will drink, one of the simplest ways of administering oral fluids is via a watering can, as in Plate 12.5. If electrolyte solutions (that is, calf scour formulations) containing bicarbonate are given, they will often stimulate closure of the oesophageal groove, transferring the fluid directly into the abomasum, where its absorption is more effective.

A faster way of administering large volumes of fluid (e.g. 10–20 litres) is by using an oral pump, as shown in Plate 12.6. The tube is protected from the cow's teeth by flexible metal rings and held in place by nose-clips ('bulldogs'). Fluids are pumped into the rumen via a stirrup pump from a bucket.

Provided that cows will drink voluntarily, it is difficult to see how additional benefit is obtained from forcible fluid administration. However, one point **is** important and that is that water should **always** be readily available to a sick cow. If she is recumbent this means offering her **warm** water five to six times daily, within easy reach. Cows will also readily drink electrolyte solutions.

Anti-inflammatory Drugs

In addition to the use of fluids, shock can also be counteracted by the use of anti-inflammatory drugs such as flunixin, phenylbutazone, aspirin or cortisone. However, these drugs are not licensed for use in dairy cows in all countries and therefore the specific instances for their use should be decided by your vet. It has been observed that cortisone reduces the inflammatory response but also allows greater bacterial multiplication. However, there is no experimental data to support this concern.

Many commercial intramammary products contain 10 mg prednisolone (a form of cortisone), which is aimed at reducing the hardness and swelling in the affected quarter. This perhaps permits better antibiotic penetration. Some cows with a typical hard quarter four to five days after a coliform infection respond well to larger doses of cortisone, either by injection or infusion into the udder. A possible serious side effect of this is that abortion may be induced.

Administration of Calcium

Some sick cows are naturally hypocalcaemic (have low blood calcium) and if dealing with a case of acute mastitis at calving, it is not always possible to be sure if a concurrent hypocalcaemia is present. Calcium is also said to aid detoxification processes in the liver. For this reason, 400 ml of calcium borogluconate, possibly mixed with glucose (as dextrose), is sometimes given to sick mastitic cows.

Administration of Glucose

Cows with acute *E. coli* mastitis can be hypoglycaemic (have low blood glucose) and may benefit from intravenous dextrose infusions (oral glucose is of no value, as it is destroyed in the rumen). In addition, the phagocytic activity of macrophages in the udder, i.e. the way in which white cells engulf bacteria (see page 23) is relatively poor, due to low oxygen concentrations in milk.

It has been suggested that the infusion of dextrose into the udder promotes phagocytic activity. Evidence supporting this is limited. Great care needs to be taken to ensure that the teat canal is not damaged and that yeasts and other organisms are not accidentally infused, making the mastitis worse.

Continual Stripping

The toxins produced by mastitis bacteria are either absorbed by the cow (possibly making her ill) or can be stripped out of the udder manually. Clearly the latter is preferable hence the importance of regular stripping of the affected quarter, maybe six or eight times daily, or more. Some people suggest stripping every 30–60 minutes until the cow is better.

The efficiency of stripping can be improved by giving oxytocin injections, which help to eject milk from the deeper parts of the gland. Natural let-down is highly unlikely to occur in a sick cow and even if the affected cow is very ill, there is likely to be a considerable amount of residual milk in the alveoli and small ducts (see page 14) and this, plus its toxins, could be removed using oxytocin.

Others have suggested leaving a strong suckler calf with the cow, to do the stripping for you! However, if the quarter is painful the cow might resent the calf sucking and only allow the normal unaffected quarters to be stripped out. In addition, the calf is less likely to suck a quarter containing bitter milk. The value of this technique is therefore a matter of conjecture.

Non-antibiotic Intramammary Infusions

The use of iodine preparations against yeast and fungal mastitis was described on page 41. Others have suggested infusing infected quarters with 20 ml of natural live yoghurt, at 12-hourly intervals for two to three days. The objectives are to decrease the raised pH of mastitic milk and to eliminate residual mastitis organisms by the probiotic effect of natural *Lactobacilli* in yoghurt. The procedure has apparently been used successfully in treating yeast and coliforms.

Topical Preparations

Products such as Cai-Pan Japanese peppermint oil ('Uddermint') have been recommended for topical application (that is to areas of the udder skin). They certainly stimulate warmth in the skin, leading to increased blood flow, but whether this can be translated into increased blood flow through the udder is difficult to assess. If such products improve the feeling of well-being in the affected cow (udder massage can be soothing) and lead to increased attention (and stripping) of the affected quarter, then they are worth using. As with so many mastitis preparations, clinical trials are difficult to carry out because a proportion of cases self-cure.

Homoeopathy

Homoeopathic medicine states that a substance which **produces** the symptoms of an illness can also be used in the treatment of any illness which causes similar symptoms. Homoeopathic remedies are all obtained from natural sources. The system relies on a series of dilutions being made, one part mother tincture to 99 parts of water and alcohol mixture. The mother tincture is an extract of natural substances, e.g. of a culture of the bacteria which originally caused the symptoms. Dilutions are repeated, and it is said that all impurities are filtered out leaving only a more potent preparation with morc energy. It is apparently this 'energy' and not the material dose of the preparation which is critical and many 'remedies' have been diluted down to well below submolecular levels.

The value of homoeopathic mastitis therapy is very questionable. At present it has the attraction of offering treatment without a milk withdrawal period. Lay homoeopathic advisers stress the need to prevent infections and promote the basic principle of sealing the teat canal between lactations, which is also the principle on which mainstream dry cow preparations are based.

Homoeopathic preparations have also been used to 'prevent' mastitis. Mixed mastitis nosodes (a nosode being a homoeopathic preparation) are administered in a variety of ways, including adding them to the drinking water, drenching and even spraying them into the vulva. Although there are plenty of anecdotal reports showing a benefit, specific trial work is lacking. Firm data will be needed before homoeopathic treatment for mastitis can be recommended.

ANTIBIOTICS IN MILK

In October 1990 the permissible level of antibiotic in bulk milk purchased from farms in England and Wales was halved to 0.005 iu/ml (5 ppb) of penicillin. In the 12 months following this reduction, the number of antibiotic failures in England and Wales doubled to almost 700 a month. From 1997, the maximum permissible level in the EC is 0.004 iu/ml.

If the normal cluster and jar are used to milk mastitic cows, they must be rinsed through with a strong solution of iodine, followed by cold water to remove all traces of chemical, between cows. Clearly by far the best procedure is to use a separate cluster for mastitic cows.

Some herdsmen do not discard milk from untreated quarters nor from cows treated with injectable antibiotics. They feel that there is little risk of antibiotic failure due to the dilution factor. This depends partly on the antibiotic in use. For example, the Delvotest (one of the most commonly used) has a specific sensitivity towards penicillin residues (see Table 12.4). Other antibiotics need to be used in higher concentrations to give a positive result.

Table 12.4 Levels at which the Delvotest gives a positive reading for different antibiotics.

Antibiotic	*Delvotest sensitivity range (ppb)*
Penicillin	2 – 4
Ampicillin	4 – 5
Cloxacillin	20 – 25
Tetracycline	200 – 400
Oxytetracycline	200 – 400
Chlortetracycline	500 – 1000
Neomycin	1000 – 2000
Streptomycin	4000 – 6000
Sulfa compounds	50000 – 100,000

A range of more specific tests which can accurately assess the levels of most antibiotics, including neomycin and sulphonamides are used in the United States. One of these, the CHARM test, is currently being used in the UK to check bulk milk prior to cheese and yoghurt manufacture. Other tests are likely to be introduced in the future to ensure that all milk is free from antibiotic residues.

A few farmers only discard milk from the mastitic quarter, despite the legal requirement to withold all milk from a treated cow. This practice is not to be encouraged. Antibiotics from the treated quarter are absorbed into the blood stream, circulate around the cow (hence the meat withholding time) and are redistributed into the untreated quarters, albeit at lower concentrations.

Table 12.5 gives **suggested** reasons for antibiotic failure. They are not necessarily the actual causes but are the most probable explanations found on discussion with farmers.

Table 12.5 A UK survey in 1980 suggested the following reasons for the presence of antibiotic in milk. (*37*)

Reason	*% of farms*
Poor or no records	32
Not withholding milk for the full period	32
Calving early/short dry period	15
Accidental transfer of milk	14
Prolonged excretion of antibiotic	12
Contamination of recorder jars	9
Withholding milk from treated quarter only	8
Lack of advice on withholding periods	6
Mechanical failure	6
Recently purchased cows	3
Milking through jars	1
Use of dry cow therapy to treat lactating cows	1

* The figures add up to more than 100% as some farms had more than one possible reason for failure.

Occasionally bulk milk fails the antibiotic test when **no antibiotics** have been used for several weeks, no cows have calved (i.e. prolonged excretion of dry cow therapy is not a possible cause) and no other cause can be identified. One possible reason for this is the presence of natural inhibitors in milk, particularly in mastitic milk. Clearly an inflamed mammary gland has the ability to counteract and kill bacteria, using natural chemicals such as lactoferrin, lysozyme and complement (see page 20).

Live bacteria are often used to test for the presence of antibiotic in milk, for example, in the Delvotest. In such tests a sample of milk is taken and incubated for four hours with a test culture of bacteria (*Bacillus stearothermophilus* in the Delvotest). By-products of bacterial fermentation produce a colour change in the milk (showing that bacteria are able to multiply) and if this occurs the sample passes the test. If antibiotic is present no bacterial growth occurs, there is no colour change and the sample has failed. Hence if milk contains a sufficient concentration of **natural** inhibitory substances, a sample could fail the antibiotic test. In fact, it has recently been shown that mastitic milk from an individual **untreated** case of coliform mastitis may fail the antibiotic test for up to 21 days after infection. Those farmers who decide not to treat clinical mastitis, but continue to put the milk into the bulk tank, could therefore still be running the risk of an antibiotic failure. It should be stressed that these results were obtained from individual animals and as yet there is no proof that the natural substances could occur in sufficient concentration to affect bulk milk. Other occasions when **individual** cows can fail the antibiotic test in the absence of antibiotics include:

- the first few days, or even weeks after calving. There is considerable cow to cow variation.
- close to drying off.
- cows with high cell counts.
- immediately before clinical mastitis (i.e. clots) is seen.

It is difficult to be precise about the interpretation of antibiotic tests on milk from individual cows or quarters, even though these tests are widely carried out. To avoid failing the test a small number of farmers dilute the milk from a suspect cow 1:5 or even 1:10 with milk from the bulk tank, before submitting the sample for antibiotic testing, which is quite permissible. This largely overcomes the problem of natural inhibitors, but antibiotic would still be detected if present in the individual animal.

The full statutory withdrawal period for an antibiotic **must** be observed, even if the cow passes the antibiotic test earlier than stated.

DRY COW THERAPY

Although opinions may vary concerning the value of treatment during lactation, there are few who would doubt the wisdom of dry cow therapy, which is the administration of a long-acting antibiotic preparation into each quarter at drying off.

Dry cow therapy is vitally important because it attempts to remove the major reservoir of contagious mastitis infection (see page 29). The advantages of dry cow therapy are:

- no milk is discarded.
- response to treatment is much more effective than that during lactation. This is shown in Table 4.2). The improved response during the dry period occurs because much higher doses of antibiotic can be infused into the quarter without having to worry about milk withholding periods. The difference is particularly apparent for *Staphylococcus aureus*, where response to lactation treatment may be especially poor.
- longer-acting antibiotic preparations can be used to improve the efficacy of action. Slow-release dry cow products are prepared by incorporating the antibiotics into waxes, by using benzathine salts or by manufacturing a product with a much smaller particle size.
- it provides some protection against summer mastitis (see Chapter 13).

Dry cow preparations should be chosen to be specifically effective against *Staphylococcus aureus* (including beta-lactamase producers) and streptococci, as these are more likely to be carried in the udder from one lactation to the next than coliforms. Table 12.6 shows the results of a survey of 294 cows (1176 quarters) found to be infected at drying off.

Table 12.6 Intramammary pathogen types and response to treatment in cows at drying off (*38*)

Bacterium	*No. of quarters*	*% of total*	*% response*
Staph. aureus	259	61	48
Other Staph types	27	6	78
Streptococci	118	28	78
Combined Staph and Strep	12	3	42
E. coli	5	1	–
Non-specific	4	1	–
Total	425	100	

Staphylococcus aureus is by far the most common organism isolated from cows at drying off. The drugs most commonly used in dry cow preparations are therefore those that are most effective against *Staphylococcus aureus*. These are:

- cloxacillin
- cephalosporins
- nafcillin
- combination pencillin/streptomycin products

Some people suggest that dry cow preparations should be changed occasionally, to avoid the development of antibiotic resistance. While the idea may be sound, provided that the correct antibiotic against *Staphylococcus aureus* is used (e.g. cloxacillin or a cephalosporin), no benefit is likely to be obtained from changing the antibiotic. Only in unusual circumstances, for example when *Pseudomonas* is a problem, are other preparations likely to be necessary.

Commercial preparations change rapidly and it is wise to consult your vet about the best product for your particular circumstances.

Frequently failure of dry cow therapy is not the fault of the drug, but rather of the way in which it is used and administered. Despite the fact that dry cow tubes contain large amounts of antibiotic, they will not necessarily eliminate bacteria accidentally infused into the udder at the time of administration as a result of poor hygiene.

Treat All Quarters

All quarters should be treated at drying off (blanket dry cow therapy) and not just those which had clinical mastitis during the previous lactation, or those with high cell counts (selective dry cow therapy). This is because:

- many cows infected during lactation never show clinical signs.
- some cows may become infected but do not show a particularly high or consistently elevated cell count. Chronic forms of *Streptococcus uberis* (see page 40) are one example of this. Even *Staphylococcus aureus* does not consistently produce high cell counts (see Table 4.3).
- every attempt should be made to prevent the establishment of chronic *Staphylococcus aureus* infections. Table 12.7 shows that response to treatment declines with age. The practical conclusion to be drawn from this is to ensure that even first lactation heifers are given dry cow therapy. The longer their udders can be kept free from *Staphylococcus aureus*, the better.

Table 12.7 Response of *Staphylococcus aureus* infections to treatment during the dry period. First and second lactation animals respond much better than older cows (*39*)

Lactation number	*Number of cows treated*	*% Response to treatment*
1–2	51	63
3–5	99	37
>5	40	33
	Total 190	Average 43

- cows are 15–20 times more likely to contract new infections in the first two weeks (and last two weeks) of the dry period. Antibiotic cover of all cows during as much of the dry period as possible is therefore likely to be highly beneficial. Because of the risk of antibiotic contamination of the milk after calving, full protection clearly cannot be given during the last two weeks of the dry period.

In one NIRD trial, in which dry cow therapy was not used, it was shown that:

- 25% of all quarters were infected at drying off.
- 5% of these quarters shed their infection naturally, i.e. underwent self-cure.
- another 10% contracted new infections

during the dry period.

- hence 30% of quarters were infected at the start of the next lactation (25 – 5 + 10 = 30).

In another trial, carried out in the Netherlands (*40*) 68 cows had only two quarters infused with dry cow therapy at drying off, their other two quarters were left untreated. During the dry period there were 10 cases of clinical mastitis in the untreated quarters, seven of which occurred in the first two weeks after drying off. Only one case of mastitis occurred in the treated quarters. This is further evidence of the benefit obtained from giving dry cow therapy to all animals.

Some say that continued use of dry cow therapy will produce such low cell counts that cows become excessively prone to *E. coli* mastitis. This is not correct. The reasons for this are given on page 26.

Administration of Dry Cow Tubes

Cows should be dried off abruptly and removed from the milking herd immediately, even if they are still giving 20–25 litres a day. If they continue to go through the parlour, milk let-down will be stimulated, i.e. the alveoli will contract, expelling milk and inhibitor protein (see page 15) thus synthesising more milk. In addition, if dry cows are left running with the main herd, there is a risk that one of them will be milked inadvertently, leading to antibiotic contamination of the bulk tank. It should not be necessary to restrict food and water severely in order to achieve drying off, although some do recommend it.

Gradual drying off is contraindicated for two reasons. Incomplete milking allows bacterial proliferation in the teat canal before therapy and hence predisposes to mastitis. In addition, cows left unmilked for 1–2 days develop large increases in cell count. In one trial (*41*), a group of late lactation cows, with an average cell count of 237,000 cells per ml were monitored. When they were left unmilked for two days the cell count increased to 540,000. After they had been unmilked for six days the average cell count increased to 5,600,000 and in one individual cow it rose to almost 17 million! Abrupt drying off is important to maintain low cell counts therefore. Similarly if a damaged teat is left unmilked for 7–14 days and allowed to heal, the first few milkings should be discarded.

A suggested procedure for drying off cows is as follows:

- dry them off in batches, making it a specific task. Split them off in the parlour at milking, then bring them back again to administer dry cow therapy. If this is done during milking, the herdsman cannot concentrate properly on what he is doing and hygiene is likely to be compromised. Dry cow therapy should also not be administered at the same time as routine foot-trimming, as the operator's hands are likely to be badly soiled and the teats will probably be splashed with faeces. Administer antibiotic first, then do the foot-trimming.
- it is **essential** to scrub the teat end with surgical spirit or commercial wipes (e.g.Mediwipes) before administration. One cause of failure of dry cow therapy is that other bacteria are introduced as it is being administered.
- only insert the nozzle a very short way into the teat or, even better, use a tube with a small, short nozzle (see Figure 12.1). Excess dilation of the teat canal produces cracks in its keratin layers, thereby compromising its defence mechanisms. In addition by squeezing the antibiotic through the teat canal, rather than fully inserting the nozzle, bacteria colonising the canal may also be killed. This may not occur if the nozzle is inserted directly into the teat sinus.
- ensure that teats are dipped immediately after tubing, thereby

removing any bacteria which might be able to colonise the teat end and produce a new infection. A few farmers regularly dip cows throughout the dry period, or at least for the final three weeks. This is excellent practice. Others suggest dipping teats after cleaning and then administering dry cow therapy through a film of teat dip.

- **record** dates of drying off and details of the tubes used. This is particularly important when a cow calves early, as an extended milk discarding period may be required.
- cows should be particularly carefully checked for mastitis in the five days after drying off and, if possible, teat-dipped daily.
- retubing. Cows which have had two or more cases of mastitis during the previous lactation are sometimes tubed again two weeks after initial administration, using a different dry cow preparation. The second product should be shorter-acting, to avoid antibiotic residues. While there is no experimental evidence to support this, it is often recommended by vets. There is, of course, a danger involved in breaking the teat seal, so meticulous attention to hygiene is required. In addition, if a second tube is used before the milk withholding time of the first tube has expired, this may lead to an overall extended withholding time. If in doubt, consult your vet.

CHAPTER THIRTEEN

Summer Mastitis

Summer mastitis has a very different aetiology and epidemiology to other forms of mastitis and does not fit either the contagious or environmental categories listed on page 29. It is essentially a disease of dry cows and heifers, although very occasionally steers, or even bulls, may be affected. The disease is common in temperate areas of the northern hemisphere, although the incidence varies enormously from one year to the next. One survey (*42*) estimated that 35–60% of herds in the UK are likely to experience the condition each year, with approximately 20,000 animals (or 1.5% of the national herd) affected. In some other countries in Northern Europe the incidence is even higher, for example in Denmark it is 5.0%. It is therefore a significant problem.

THE BACTERIA INVOLVED

A total of six organisms have been isolated:

- *Actinomyces (Corynebacterium) pyogenes*: this is the most frequent isolate and is the organism responsible for the severe necrosis and destruction of the quarter.
- *Peptococcus indolicus*: ferments milk and damaged tissue into organic acids and indole and is responsible for the characteristic foul smell.
- *Streptococcus dysgalactiae*: this may be the primary infection, allowing *Actinomyces pyogenes* to enter and/or proliferate in the mammary gland. It is commonly found on flies and on damaged teat skin.
- Microaerophilic cocci: sometimes known as Stewart-Schwann cocci.
- *Bacteroides melaninogenicus*
- *Fusobacterium necrophorum*

However, by no means are all six organisms isolated from every case of summer mastitis. Table 13.1 shows the percentage of occasions on which each organism is isolated. *Actinomyces pyogenes* and *Peptococcus indolicus* are the most commonly isolated in the UK, whereas in Denmark the Stewart-Schwann coccus is more common.

Table 13.1 The percentage of occasions on which different bacteria are isolated in summer mastitis cases, in various European countries (*43*)

	% isolated		
Bacterium	*UK*	*Denmark*	*Holland*
A. pyogenes	85	72	37
P. indolicus	62	87	33
S. dysgalactiae	24	37	8
S-Schwann coccus	22	83	–
F. necrophorum	1	51	22
B. melaninogenicus	<1	35	8

MODE OF TRANSMISSION

The major means of transmission of infection in the UK is thought to be by the sheep head fly, *Hydrotoea irritans*, which lives by sucking the blood of cattle. The fly prefers woods, copses and damp ground which is sheltered from the wind. Larvae overwintering in light, sandy soils emerge as adults in July. They are present primarily during July, August and September and these are therefore the most common months for summer mastitis. Cases may also occur in June and October, if the weather is exceptionally hot and humid. Eggs are laid into the soil in October, and there is only one generation of adults each year.

The flies live in bushes and trees and only fly out to feed on cattle when wind speeds are low (less than 20km per hour) and in the absence of rain. Their favoured landing areas are the legs, abdomen and udder. Fore teats are more commonly affected than rear, possibly because the swishing tail removes flies from the hind teats.

Although there is considerable evidence that *Hydrotoea irritans* is a vector in the cause of summer mastitis (and hence fly control is a major part of prevention), there are still doubts about it being the only factor involved. This is because:

- Hydrotoea flies are often found in association with cattle, but without causing summer mastitis. Possibly some other factor simultaneous with the presence of the fly is required to damage the teat end: for example thorns, nettles, thistles or long grass, another type of fly, or even cattle licking themselves excessively.
- summer mastitis can occur in parts of the world where *Hydrotoea irritans* is not present.
- disease can occur in winter (usually associated with teat-end damage) when there are no flies.
- although many of the bacteria causing summer mastitis can be found in the intestine of the fly and are regurgitated during feeding, experiments which attempt to transmit summer mastitis from infected flies to cows have been unsuccessful. Experimentally it is possible to induce summer mastitis by infusing *Actinomyces pyogenes* and *Peptococcus indolicus* through the teat canal.

One theory therefore is that the first case of summer mastitis occurs spontaneously, possibly by infection tracking in from infected teat sores, and subsequent cases are caused by flies spreading the infection. Outbreaks of disease do occur and hence there must be some vector, perhaps in association with a reduction in the immune status of the animal.

CLINICAL SIGNS

The classic symptoms of summer mastitis are a hot, hard and swollen quarter, usually with a tense and enlarged teat, as in Plate 13.1. The quarter is painful and the secretion thick and clotted, with a characteristic foul smell. More severely affected cows have a raised temperature, are lame because of the painful quarter and may develop swollen hocks. Some animals may abort (summer mastitis primarily affects pregnant cattle), and others give birth to a full-term but retarded and weakly calf. Neglected cases can easily die and as dry

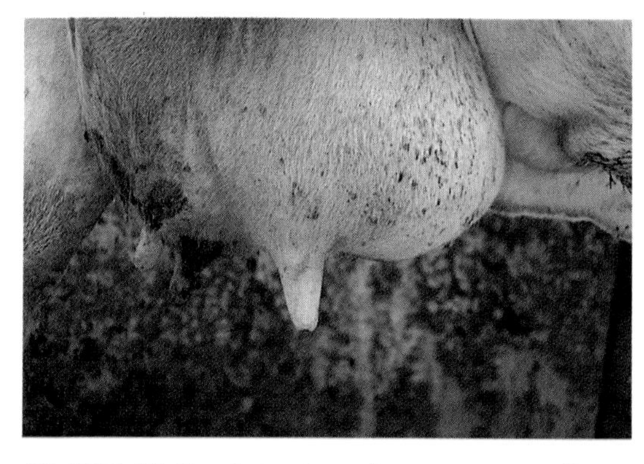

PLATE 13.1 A teat and quarter swollen with summer mastitis. Sometimes the legs also swell.

cows and pregnant heifers sometimes do not receive as much attention as they should, death is not uncommon. Prompt treatment is certainly important.

In some cows and heifers, the disease is very mild and is not seen during the dry period. It is only after calving, when a blind quarter is detected, that previous infection becomes apparent. These animals have a thickened teat, with a fibrous core running down the centre, replacing the teat cistern. Roll the teat between your finger and thumb and compare the feel with an adjacent non-affected teat to emphasise the difference. Attempts to infuse antibiotics demonstrate how small the cistern has become – much of the antibiotic will run back out under pressure.

A further syndrome, which has increased in frequency over the past few years, is seen in cows calving down with what appears to be a low-grade mastitis, with just a few clots at each milking. On culture, this proves to be summer mastitis (*Actinomyces pyogenes*). Presumably only a very small part of the mammary gland is affected and it is only when the gland becomes active, as at calving, that clinical signs appear. However, these cows often do not recover, even with high and prolonged doses of antibiotic.

TREATMENT

The two main organisms causing summer mastitis (*Actinomyces pyogenes* and *Streptococcus dysgalactiae*) are both highly sensitive to penicillin and hence penicillin and its derivatives are the antibiotics of choice for treatment. Even so, very few quarters ever recover. Intramammary tubes are of very doubtful value.

If at all possible, the infected teat should be stripped very regularly, especially during the first two to three days. This will then reduce the chances of an abscess bursting through the side of the udder, as seen in Plate 13.2. However, some animals strongly resent manual stripping and an alternative is to drain the udder by making a longitudinal cut through the teat, as in Plate 13.3. (An anaesthetic is required and it is definitely a job for the vet). Infection drips from the teat onto the ground. However, as the environment is already highly contaminated with *Actinomyces pyogenes* (which is a normal environmental organism), provided the cow is removed from the susceptible group, to avoid

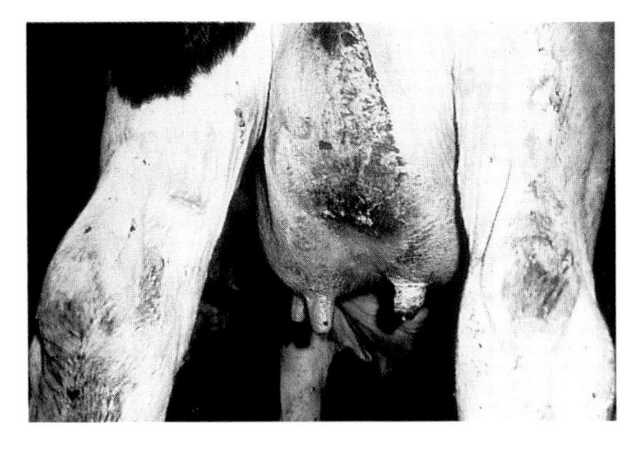

PLATE 13.2 Summer mastitis. When it is neither practical nor possible to strip the teat, the udder may burst and discharge.

PLATE 13.3 When manual stripping is not possible, the teat can be drained by making a longitudinal slit. However, the cow will still be discharging infection and needs to be removed from other dry cows.

fly-borne transmission, the risk is minimal. Even if the teat is not opened, the affected animal should **always** be removed from the rest of the group to avoid spreading infection to other cows.

CONTROL

Prevention of summer mastitis is primarily based on fly control and antibiotic therapy. The most common measures are:

Reduce Exposure

Keep susceptible cattle away from known summer mastitis pastures. Exposed, open fields on high ground, with a clay soil, are ideal as the sheep head fly dislikes these conditions. Avoiding high risk areas is probably the best control measure.

Fly Control

There are a variety of methods available, but most rely on the flow of sebum over the body surface. However, the udder has no sebaceous glands and so there is no flow of sebum over teat skin (see page 12).

- fly tags give good protection of the head and back, particularly if two are used, one in each ear. The abdomen and udder, are still not well protected, however, and these are the favourite landing places for *Hydrotoea irritans*.
- pour-on preparations are applied along the animal's back and also give poor teat protection. In addition, during wet periods, when flies are most active, their persistence is reduced.
- sprays: one needs to be very conscientious to achieve a thorough covering of each animal, including the udder.
- micropore tape. Sealing the teat ends with tape has been used successfully on the Continent, particularly in Denmark, but is not popular in the UK. It is not easy to apply and has to be replaced every three weeks.
- by far the best approach is to apply fly repellent directly onto the udder and teats **every week** in high risk areas. Although this incurs a huge labour cost, only one animal has to be saved to make the effort worthwhile. In very high risk areas some farms successfully use weekly applications of a mixture of pour-on fly repellent and Stockholm tar, although such preparations are of course unlicensed. Stockholm tar alone, regularly applied to teats, is also effective. Chlorhexidine teat dips combined with a fly repellent are available, although to be effective they must be applied daily.
- segregate and house affected animals: this removes an important source of infection.

Dry Cow Management

This undoubtedly helps, as most cases occur four or more weeks after drying off, in other words, when the antibiotic concentrations are declining.

- in high risk areas, repeat infusions of dry cow antibiotic after three weeks are advisable. Careful attention to expected calving dates is needed to avoid antibiotic contamination of milk after calving. Consult your vet to decide which product to use.
- heifers can be tubed, but the tip of the tube must be abutted **against** the teat orifice and the antibiotic squeezed through the teat canal under pressure, rather than inserting the tube into the teat canal itself.
- house the late pregnant dry cows: *Hydrotoea irritans* will not enter buildings and therefore there is less irritation and nuisance inside. The cows can go out later at night, i.e. after dark, when the fly is not active.
- move the calving pattern to earlier in the summer, so that there are fewer dry cows in July, August and September.

- on some small farms the dry cows are run with the milkers so that they can be teat-dipped and a watchful eye kept on the udder. Provided the dry cows are clearly identified (to avoid being milked) this form of control can prove to be very effective.

As approximately 20,000 animals are affected in the UK each year, summer mastitis continues to represent a major cost to the dairy industry. It seems extraordinary that we do not have adequate technology to control flies.

CHAPTER FOURTEEN

Diseases of the Udder and Teat

The herdsman will encounter a wide range of diseases affecting the udder and teats. These are not always directly related to mastitis, but any problem involving the mammary gland has the potential of increasing susceptibility to mastitis. In addition, a few of these diseases may be confused with mastitis. Some of the more common conditions are described in the following section.

DISEASES CAUSED BY METABOLIC DISORDERS

Blood in Milk

This is seen in freshly calved cows and can vary from a few clots of blood in the milk from one quarter (Plate 14.1) to almost pure blood coming from all four quarters (Plate 14.2). Some herds may experience a very high incidence in freshly calved cows, resulting in large quantities of milk being discarded. Although extensively investigated, often no cause is found. In individual animals, blood in milk may be the result of trauma at calving (the legs bruising the udder during uterine contractions), excessive udder oedema, cows with unusual gaits, or pendulous udders which are knocked by the legs when walking. In occasional cases, rupture of the anterior udder ligaments (see page 8 and Plate 2.3) can produce severe blood in the milk and the blood may even discharge through the ruptured skin at

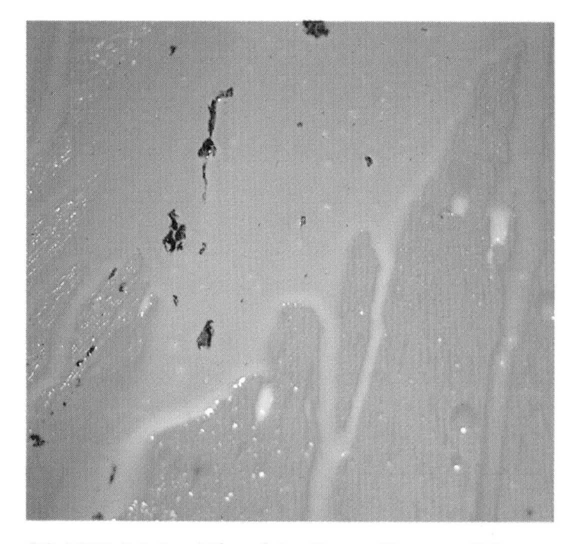

PLATE 14.1 Blood in the milk: a mild case.

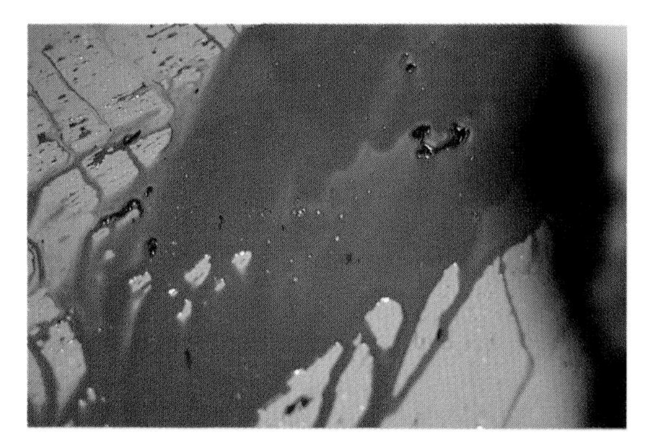

PLATE 14.2 Blood in the milk: a severe case, with almost neat blood drawn from the udder.

the front of the udder.

There is no useful treatment for blood in milk, although most people recommend only light milking (sufficient to flush bacteria from the teat canal), thus producing increased pressure within the udder, in an attempt to stop the bleeding.

Necrotic Dermatitis/Intertrigo

This is often first noticed because of its purulent smell! A foul, moist discharging area is seen at the front of the udder (Plate 14.3). The condition occurs primarily in freshly calved cows, especially older animals, and is thought to be caused by an ischaemic necrosis (death of tissue due to lack of blood) of the skin. Severe congestion of the udder around the time of calving is thought to lead to stasis of blood and subsequent degeneration of the tissues.

No treatment is particularly effective, but thoroughly washing the area with antiseptic, removing dead tissue and applying glycerine or an antiseptic ointment will help.

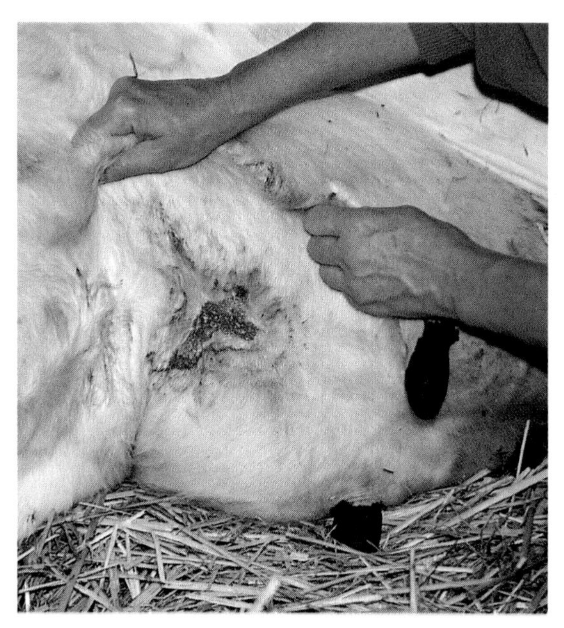

PLATE 14.3 Necrotic dermatitis: a foul-smelling sore at the front of the udder.

'Pea' in Teat

The first sign of a 'pea' in the teat will be when one quarter is found to be full of milk after the cluster has been removed. Hand-stripping will initially produce a few good draws, but flow suddenly stops. At this stage the 'pea', a thick pad of fibrous material, has become lodged in the teat canal (Plate 14.4). Enlargement of the teat canal under local anaesthetic is one method of removal. A variety of shapes, sizes and colours of 'pea' is seen (Plate 14.5). They occur most commonly in freshly calved cows, usually up to peak yield and, from their red colour, must originate from blood clots.

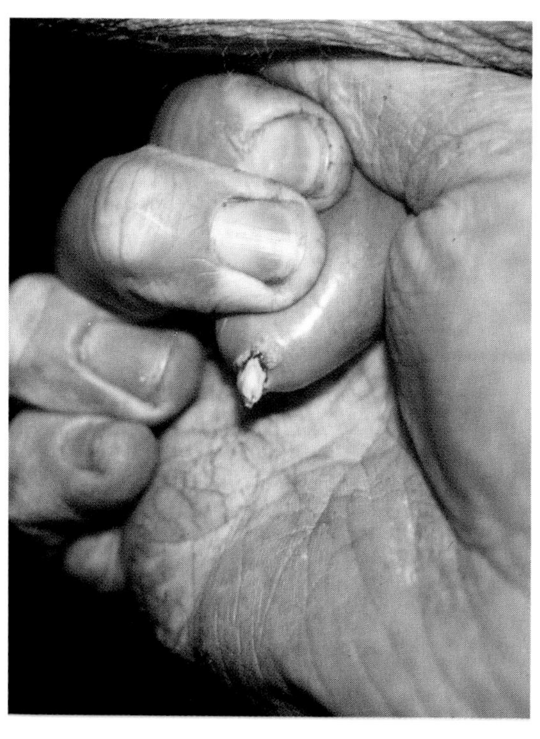

PLATE 14.4 'Pea' protruding from teat.

PLATE 14.5 'Peas' are found in a variety of shapes and colours.

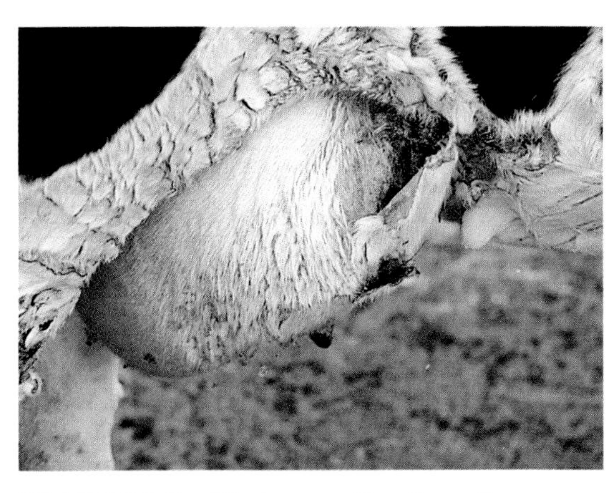

PLATE 14.6 Photosensitisation: the udder skin thickens, becomes dry and in the later stages peels off.

Photosensitisation

Occasionally photoreactive chemicals accumulate under the skin of individual cows. These are chemicals which react with sunlight and ultraviolet rays. When exposed to ultraviolet light the chemicals produce thermal energy, which in turn causes intense inflammation, very similar to a burn. Only white or lightly pigmented skin is affected, since black skin prevents absorption of ultraviolet light.

The initial photoreactive agents may have been eaten (e.g. St. John's wort in the UK, or lantana poisoning in New Zealand), or may be produced as a result of liver damage. The teat skin is initially thickened and often very painful. It later becomes dry and peels off, leaving a raw surface beneath (Plate 14.6) before eventually healing.

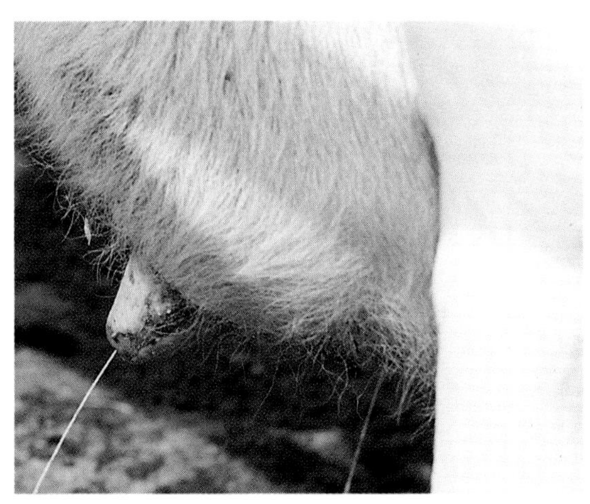

COURTESY OF PROFESSOR D. WEAVER

PLATE 14.7 Teat sunburn: note it affects one side of the teat only.

Teat Sunburn

Occasionally cows with non-pigmented teats and large udders develop sunburn along one side of the teat (Plate 14.7). This can be an irritant and if flies are attracted, may develop into a summer sore. The use of emollients and fly repellents is effective.

Udder Oedema

'Oedema' is the name given to an accumulation of fluid in and under the skin. The classic test for oedema is to press the surface of the udder with your finger for four to five seconds: a pit remaining at the point where you applied pressure (Plate 14.8) is characteristic of

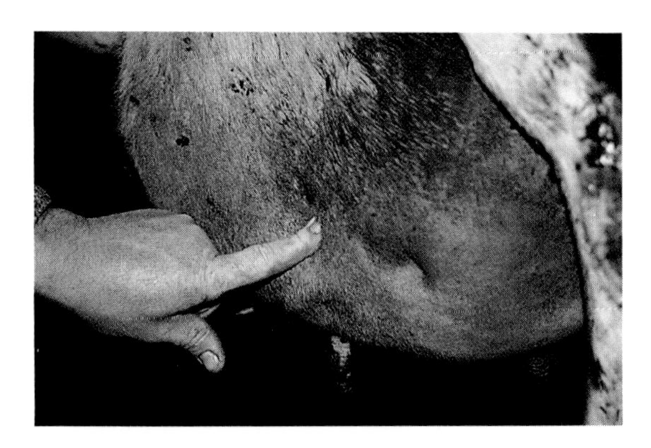

PLATE 14.8 Udder oedema characteristically leaves a 'pit' after finger pressure.

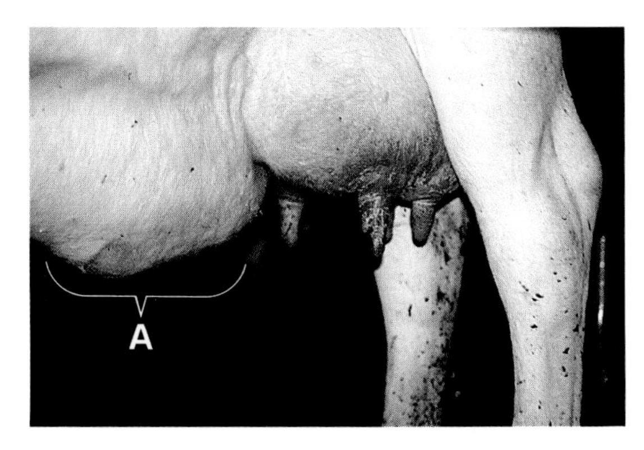

PLATE 14.9 Oedema under the belly (A). Note also the dry, cracked teat skin (teat necrosis).

oedema. It can also occur on the lower abdomen, running from the front of the udder towards the forelegs (Plate 14.9). Note how in the heifer shown the teat skin is dry and cracking, again due to poor circulation. This is the early stage of necrotic dermatitis.

Excess udder oedema can become a problem, particularly in heifers. In its most severe form it can lead to such extensive necrotic dermatitis that the teat and udder skin eventually sloughs (i.e. falls off). These animals are impossible to milk and have to be culled. Many develop mastitis. Even in those which **can** be milked, milking is such a painful process that let-down is poor and yields suffer. Because of the turgidity of the teats there is an increased incidence of liner slip and teat-end impacts, which will increase the risk of mastitis (see page 63). Finally, gross congestion of the udder puts an enormous strain on the suspensory ligaments, which may then rupture (see page 8), seriously reducing the longevity of the heifer.

Possible causes of excess udder oedema at calving include:

- overfat heifers.
- excessive feeding immediately prior to calving.
- overzealous pre-calving mineral supplementation, leading to fluid retention. The author has seen udder oedema problems resolved coincident with the removal of ad lib. mineral supplementation. Feeding caustic-treated straw or wheat has also been suggested as a causal factor, because it leads to excessive sodium intakes.
- inadequate exercise. Natural flow of fluid from the udder is via the lymphatics (the body fluid drainage system), moving upwards towards the pelvis. The flow of lymphatic fluid is promoted by limb movement during exercise. Lack of exercise at calving increases fluid stasis, leading to oedema.

Over the past few years a new form of udder oedema has begun to increase in incidence. This is sudden in onset, may affect one or perhaps two quarters only and may be seen in cows in mid-lactation, well after the periparturient oedema (close to calving) has disappeared. The cause is unknown. Affected animals often respond slowly to diuretics i.e. drugs which remove excess fluid from the body. Cows may be difficult to milk while the condition is present. This is because the teat may almost disappear into the hard, swollen and oedematous quarter. On first sight the herdsman is highly likely to suspect mastitis, but there are no changes in the milk, there is no increase in body temperature and the cow is not off colour in any way. The finger-pressure test shows that the swelling is a typical oedema.

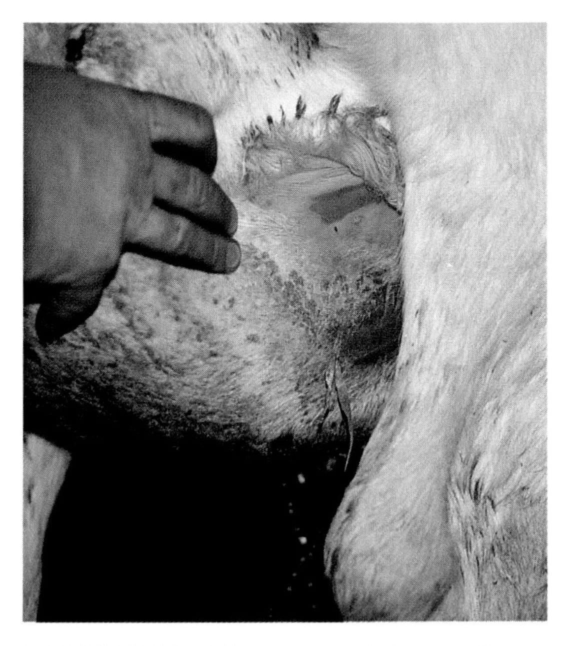

PLATE 14.10 Wet eczema between the legs and udder, seen mainly in heifers.

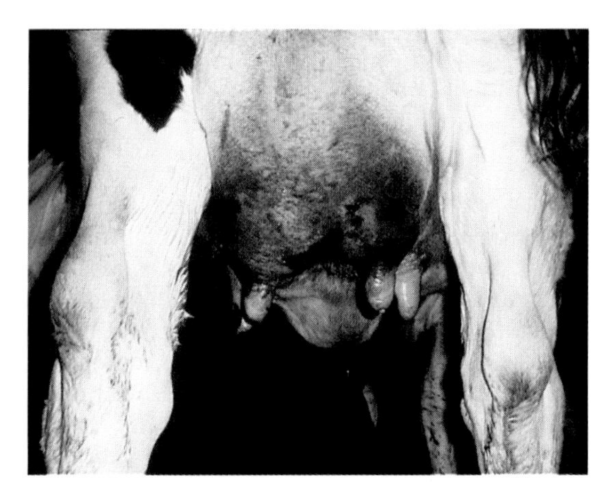

PLATE 14.11 Advanced wet eczema which has developed into necrotic dermatitis: some heifers are so badly affected that they are impossible to milk.

Wet Eczema in Heifers

This is also thought to be a degeneration of skin caused by excessive udder oedema and is most commonly seen between the legs and udder (Plate 14.10). More advanced cases may develop into a necrotic dermatitis affecting the whole udder (Plate 14.11). The skin is initially swollen and thickened, later becoming dry with a flaking surface. Occasionally heifers are so badly affected that they become impossible to milk. In other cases damage to the teat end leads to mastitis.

DISEASES WITH INFECTIOUS CAUSES

Bacterial Eczema

A relatively uncommon form of teat eczema is shown in Plate 14.12. Note how only one side of the teats is affected. This was caused by an open sore on the lower lip of this beef suckler cow (Plate 14.13) and hence the teats were only affected on the side of the sore. A good response was obtained to parenteral (injectable) antibiotics and topical antiseptic teat ointment. The most probable cause of the lesion was *Fusobacterium necrophorum*, although a culture was not carried out to confirm this. The same organism causes 'blackspot' on the teat end.

Bovine Herpes Mammillitis

This is a much more serious viral infection of the teats than pseudocowpox, and in some cases can lead to such severe and painful teat skin damage that the animal becomes impossible to milk. In appearance it is very similar to necrotic dermatitis (seen in Plates 14.11 and 14.3), but usually with the teats more affected than the udder. Treatment

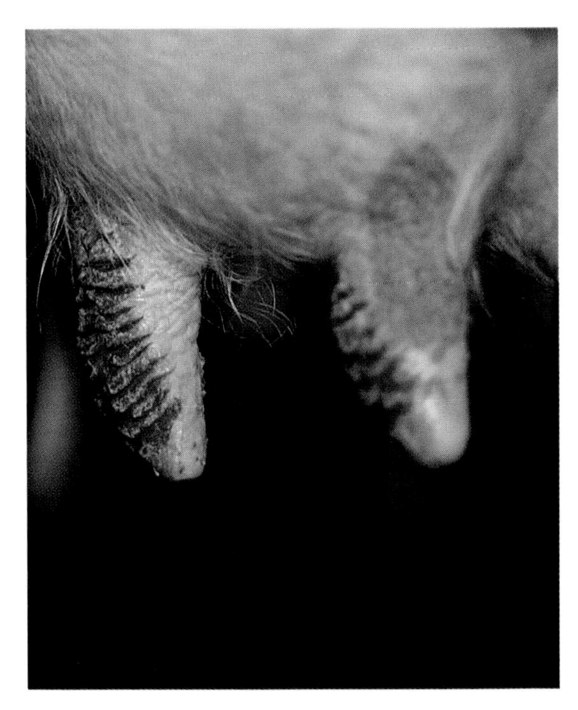

PLATE 14.12 Bacterial eczema: this case was caused by contact with an infected sore on the cow's lower lip. Note that eczema appears only on one side of the teats.

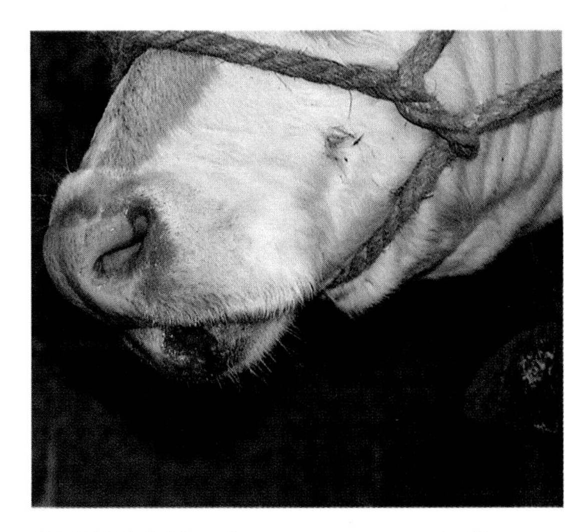

PLATE 14.13 Open sore on cow's lip, acting as the source of infection for the teat lesion in plate 14.12.

with emollient dips helps, but teats are slow to heal. In painful cases, predipping with glycerine (which should be wiped off prior to the application of the cluster) will soften the teats and assist milking. During the active phase of the disease, the vesicles (fluid blisters) which appear on the teat skin contain large numbers of virus particles. Affected cows should therefore either be milked last or the milking unit should be thoroughly cleaned and disinfected between cows. Fortunately, once a cow has been infected and recovered, she is left with life-long immunity. The condition is virtually never seen in dry cows and some people consider that herpes mammillitis virus may remain dormant on carrier dry cows to become active and cause disease after calving.

Pseudocowpox

This is a viral infection of teat skin and produces characteristic horseshoe-shaped lesions. The teat shown in Plate 14.14 is quite extensively affected. More commonly a smaller area of teat skin with a smaller and less well defined lesion is seen, as in Plate 14.15. In the initial stages there is commonly redness of the skin which develops into pustules and finally forms scabby areas which when removed expose the horseshoe-shaped lesions. The condition is not particularly painful and milking can continue. Most animals heal in 3–4 weeks, resolution being assisted by the use of teat dips containing an emollient. Provided that the weather conditions are mild, hypochlorite dips are thought to be particularly effective. Iodine dips may also be used. Dips are probably better than sprays, as they achieve a more thorough cover. They also reduce the growth of mastitis bacteria such as *Staphylococcus aureus* and *Streptococcus dysgalactiae* which could otherwise proliferate in the pseudocowpox scars. Greasy ointments are not recommended, as they attract dirt, may spread bacteria and do not kill viruses.

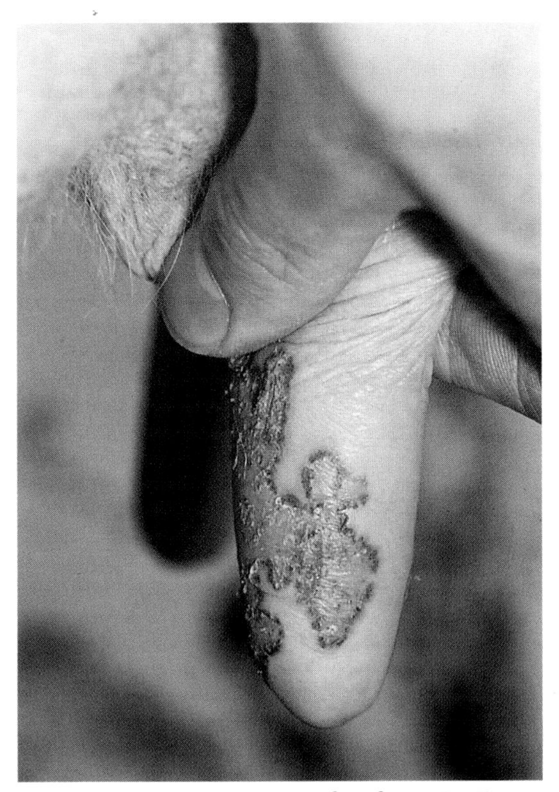

Courtesy of Professor D. Weaver

PLATE 14.14 Pseudocowpox: characteristic spreading area of superficial, non-painful haemorrhage.

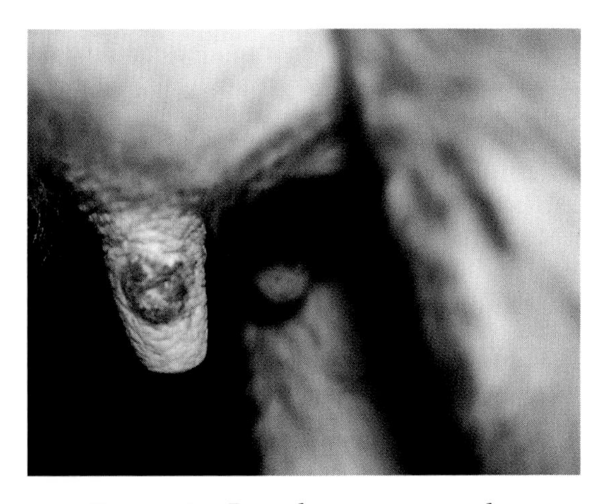

PLATE 14.15 Pseudocowpox: single circular lesion – this is the most commonly seen form.

Immunity to pseudocowpox is short-lived and further infections can occur 6–12 months later. The same virus may also produce small warts, sometimes called **milker's nodules**, on the herdsman's hands. It has been suggested that pseudocowpox is related to orf, because it is often seen in herds which have contact with sheep. However, the milker's nodules seen in man are very different from human orf lesions.

Staphylococcal Impetigo

Not a common condition, this is a red, raw rash, appearing on the surface of the udder (Plate 14.16). The lesions are not particularly painful, but they produce moist, pimply areas on the udder skin and could represent a reservoir of mastitic bacteria. Washing the skin and treatment with topical antiseptics is usually effective.

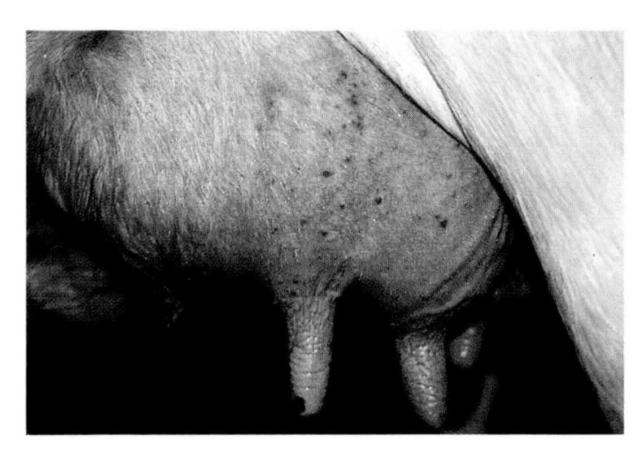

PLATE 14.16 Staphylococcal impetigo. Note the red rash on the udder.

Summer Sores (Licking Eczema)

This is thought to be caused by fly irritation. Some cows lick their teats and abdomen excessively, causing surface erosions and sores. A typical example is seen in Plate 14.17. Much worse cases can occur and may lead to summer mastitis. Treatment with fly repellents allows rapid healing. Teat dips may also help.

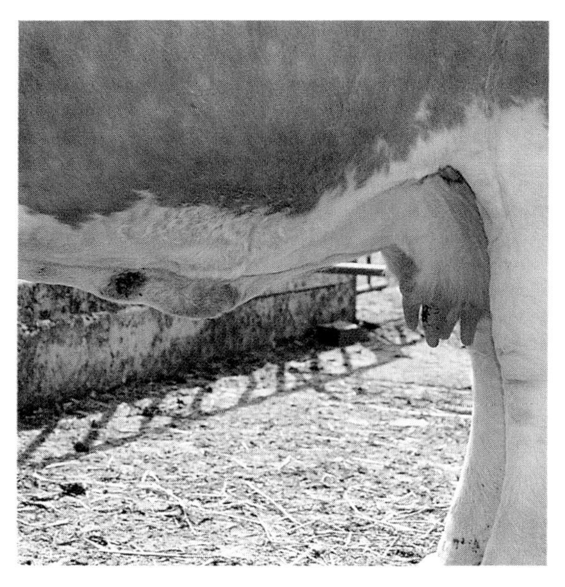

PLATE 14.17 Summer sores: fly irritation and subsequent excess licking is the probable cause of these belly and teat sores.

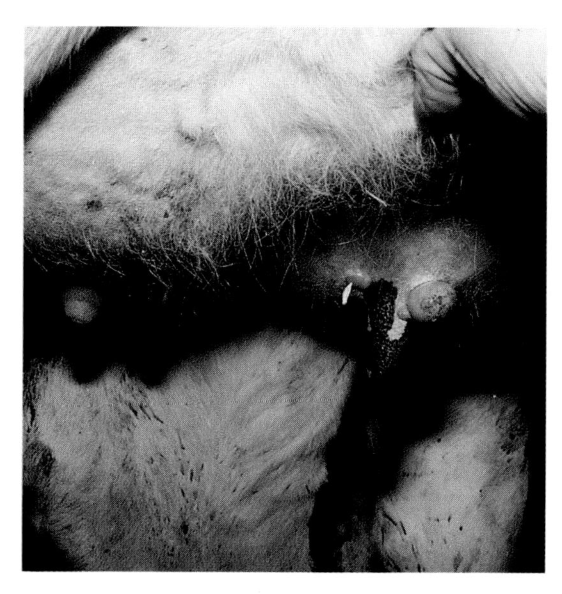

PLATE 14.18 'Fleshy' teat warts.

Teat Warts

Warts are caused by papovaviruses. There are five different strains of virus, which possibly explains the big variation in the type of wart seen. The most common are fleshy nodules (Plate 14.18) and feathery warts (Plate 14.19). The latter can be pulled off quite easily, as their root readily detaches. Nodular warts are more difficult to remove.

Vaccines have been prepared by grinding up the wart to release the virus, inactivating it with formalin and then injecting the filtrate back into affected animals. A licence may be required to do this. Such vaccines only seem to be of limited value as there is often a poor response to treatment. Most animals eventually undergo self-cure and by the second or third lactation the warts have gone.

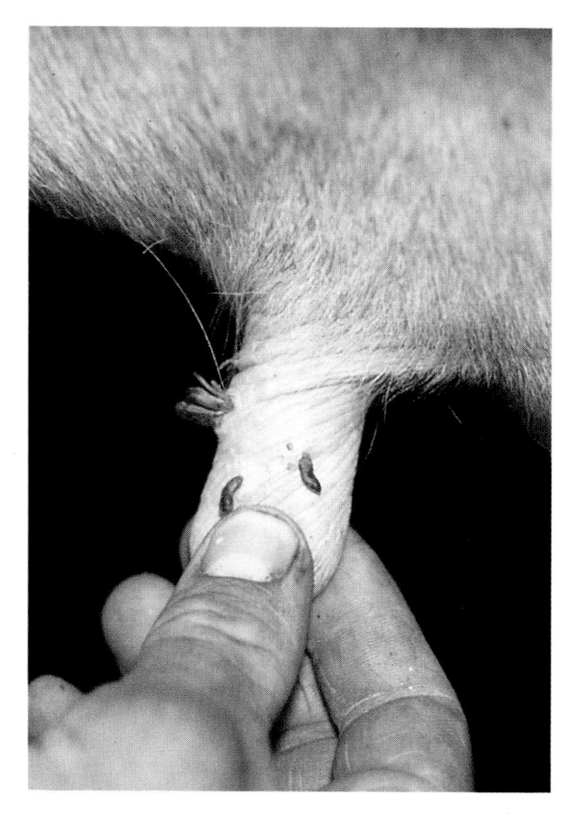

PLATE 14.19 'Feathery' teat warts.

When present on the teats they can cause considerable disturbance to milking:

- they may lead to poor liner attachment, air leakage and therefore teat-end impacts (see page 63). In some heifers the warts may be so extensive that the animal is impossible to milk.
- warts may be painful, thereby inhibiting milk let-down, increasing residual milk and decreasing overall yields (see page 90).
- warts around the teat canal can predispose to mastitis.
- skin damage from warts could predispose to secondary infections with *Staphylococcus aureus* and *Streptococcus dysgalactiae.*

The virus causing warts is thought to be transmitted by flies and certainly warts are seen more commonly in heifers reared near to rivers and streams (an ideal habitat for flies). Fly control is therefore an important preventive measure. This is discussed on page 166.

However, flies are not the only vector for transmission, since warts may also be seen in housed heifers, especially when stocking densities are high.

DISEASES CAUSED BY PHYSICAL TRAUMA

Causes of Teat Injuries

Some herds seem to suffer almost epidemics of teat crushing and teat injury. Factors to consider as possible causes when a high incidence is encountered include:

- high stocking densities and inadequate loafing areas: cows are simply too tightly packed together.
- slippery floors and passageways. Dramatic improvements are often seen following grooving of the concrete, to provide a surface with a better grip. (Heat detection may also improve!)
- excessively narrow cubicle passages: cows either reverse clumsily into the cubicle opposite, or may fall when pushing past one another.
- very narrow cubicles: large cows may push their legs through into the adjacent cubicle and damage the teats of their neighbours.
- poor cubicle comfort, for whatever reason, could lead to an increased number of cows lying outside on slippery surfaces and hence increase teat injuries. Cubicle design and dimensions are discussed on page 107.
- slippery cubicle beds, for example rubber mats with inadequate bedding: if the bed surface is too smooth, a cow may fall and injure her teats while attempting to stand.
- loose housing, particularly in long, narrow and poorly designed yards and where cows are heavily stocked: although cubicle systems can produce teat injuries, they are probably more common in poor loose yard systems.
- rough handling, such as rushing the cows along passageways and around corners, so that they are liable to fall.
- continually changing groups: once cows have settled into a group of 50–100 animals, they are best left as such. Moving animals from one group to another leads to aggression and fighting and could produce teat injuries.
- inadequate fly protection: cows grazing outside, irritated by flies may chase around fields and through fences, injuring their teats. Dogs could conceivably produce a similar effect. However, most teat injuries occur in housed cattle.
- increased lameness: cows feeling uncomfortable on their feet are stiff and cumbersome when rising and are likely to have an increased incidence of self-inflicted injuries.
- poorly maintained buildings: jagged edges, especially on cubicle beds, could increase the incidence of teat damage.

Amputation of Teats

It is surprising how many cows arrive for milking having totally amputated one of their teats at its base, as in Plate 14.20. Many of these cows continue to produce milk in all four quarters, the affected gland simply discharging onto the parlour floor when let-down occurs. Unfortunately the cow in Plate 14.20 developed a severe mastitis and had to be culled.

PLATE 14.20 Teat accidentally amputated at its base – this is a surprisingly frequent occurrence.

Blackspot

This is the term given to a necrosis of the teat sphincter, often with a secondary bacterial infection with organisms such as *Fusobacterium necrophorum.* A typical example is shown in Plate 14.21. Because of the extensive teat-end damage, the risk of mastitis is enormous. Affected cows are best hand-stripped for one or two weeks and allowed to heal. The use of a chemical debraiding ointment improves the rate of healing.

Blackspot may initially be the result of machine damage to the teat end, followed by exposure of the teat to an adverse environment, ie. dirty conditions. Low-emollient teat dips may exacerbate this, although there is some anecdotal evidence that hypochlorite dips are beneficial, in that they promote healing by removing dead tissue from the teat end.

PLATE 14.21 Blackspot, an infected erosion of the teat end.

Chemical Teat Damage

The most common mistake is to accidentally use an iodophor/phosphoric acid bulk tank cleaner as a teat dip. This has happened on numerous occasions and can lead to severe problems with chapped teats, sores and subsequent mastitis. It can also affect the milker's skin. Drums of chemicals **must** be carefully labelled.

Crushed Teats

Plate 14.22 is a typical example of a cow whose teat has been crushed, rendering her extremely difficult to milk. This photograph was taken immediately after

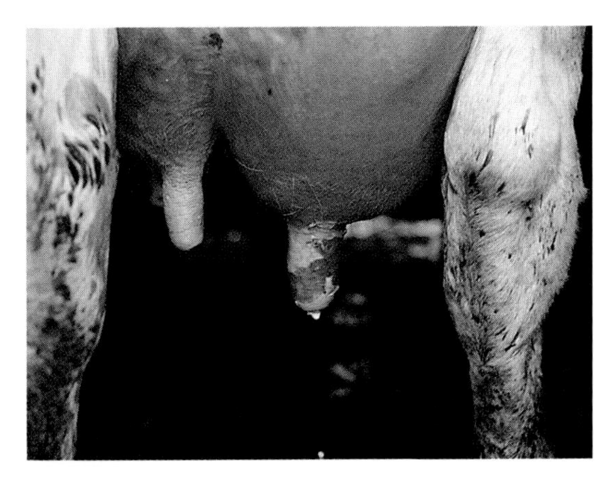

PLATE 14.22 Crushed teat engorged with milk because the cow could not be milked.

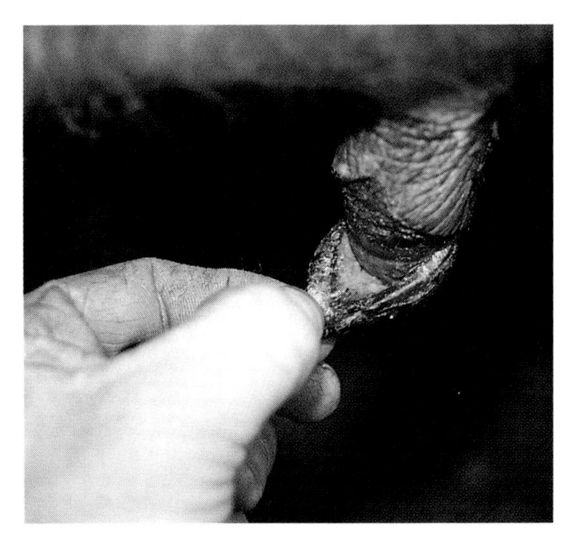

PLATE 14.23 Typical cut teat before removal of skin flap.

milking and demonstrates that the affected quarter has not milked out properly. The preferred course of action by many herdsmen is to simply stop milking the teat until it has healed. Within a few days the pressure of milk within the udder declines and, by not applying the milking machine, healing occurs much more rapidly. Admittedly there is a risk of mastitis, although this is no greater (and is probably less) than if a teat cannula is inserted and left in position for one or two weeks. If the cow will permit it, the risk can be minimised by a few hand-strippings at each milking.

When milking is resumed, the quarter returns to production surprisingly quickly, even if it has not been milked for three or four weeks. However, the first few milkings should be discarded, as this milk will have a very high cell count (see page 161).

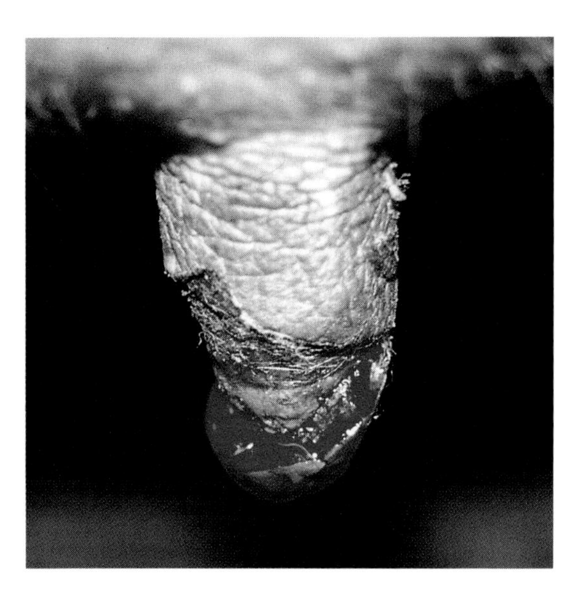

PLATE 14.24 Amputation of skin flap promotes healing.

Cut Teats

Teats are subjected to a wide variety of cuts and lacerations, one of the most common being a horizontal cut on the lower third and towards the teat end, as seen in Plate 14.23. Although this cut has not penetrated through to the canal, the cow will be difficult to milk because the flap of skin will be pulled down each time the unit is pulled off. It is unlikely that the skin flap will be thick enough for successful suturing.

The most effective treatment is to

remove the flap under local anaesthetic (as in Plate 14.24) and perhaps leave the teat to heal for a few days before starting to milk it again. Most of these cuts heal extremely well. During the early stages the wound can be protected from dirt and flies with Micropore tape, a thin bandage which allows the wound to 'breathe', thereby promoting healing.

Hyperkeratosis (Sphincter Eversion)

Hyperkeratosis of the teat end is seen as a protrusion of dry, creamy-white tissue surrounding the teat sphincter. This condition is also known as sphincter eversion. A typical example is seen in Plate 14.25.

A degree of hyperkeratosis may be a normal feature of high-yielding cows, as it is seen particularly at, or soon after, peak lactation. The keratin lining the teat canal is thought to 'trap' bacteria by adhesion. This keratin is then flushed from the teat during milking. Appearance of excess keratin at the teat orifice is referred to as hyperkeratosis. More severe hyperkeratosis is an abnormality. It may be associated with adverse milking machine function, excessive vacuum fluctuations (note the haemorrhage on the teat end in Plate 14.27) and overfierce pulsation, leading to total stripping of keratin from the canal. This can remove defence mechanisms and will predispose to mastitis. Incomplete or slow opening of the liner (namely the 'A' phase of the pulsation cycle, see page 53) is an example of how poor pulsation can affect hyperkeratosis. If milk flow starts while the liner is still partially closed, then milk is effectively being 'squeezed' out through the teat end. Keratin is dragged out with the milk and hyperkeratosis results. This may happen with old liners. Old liners open more slowly due to partial collapse of the rubber. They are therefore likely to open at a higher vacuum level, and this could lead to hyperkeratosis.

Teat-end Scores

Attempts have been made to devise a teat scoring system based on the appearance of the teat orifice. On a 0–4 basis this is approximately:

- 0 – the perfect teat end. Although there may be a slightly thickened ring visible (the teat sphincter), there is no roughening.
- 1 – the orifice appears slightly too 'open' and has lost its normal smooth, circular appearance.
- 2 – moderate hyperkeratosis: a few small, rough fronds of keratin are protruding 1–2 mm from the raised teat orifice (Plate 14.25).
- 3 – orifice very rough, with keratin protruding all the way round the teat spincter.
- 4 – advanced protrusion of keratin, 2–4 mm long, with sphincter giving the appearance of having turned almost inside out (Plate 14.26).

Only scores of 3 and 4 are likely to lead to an increased incidence of mastitis.

Teat Chaps

'Chaps' are cracks in teat skin. They occur particularly when cows are exposed to wet, cold and windy weather, or to damp and dirty environmental conditions. Teat dipping in severe cold, for example in subzero conditions can produce chaps, especially if damp teats are exposed to a wind chill factor. Their development may be aggravated by poor unit alignment, leading to twisting of the teats (and hence opening of skin cracks) when the cluster is removed. Post-milking disinfection with high emollient dips, or even neat glycerine, promotes rapid healing.

Not only are chaps painful, but they can also harbour mastitic bacteria, particularly *Staphylococcus aureus* and *Streptococcus dysgalactiae.*

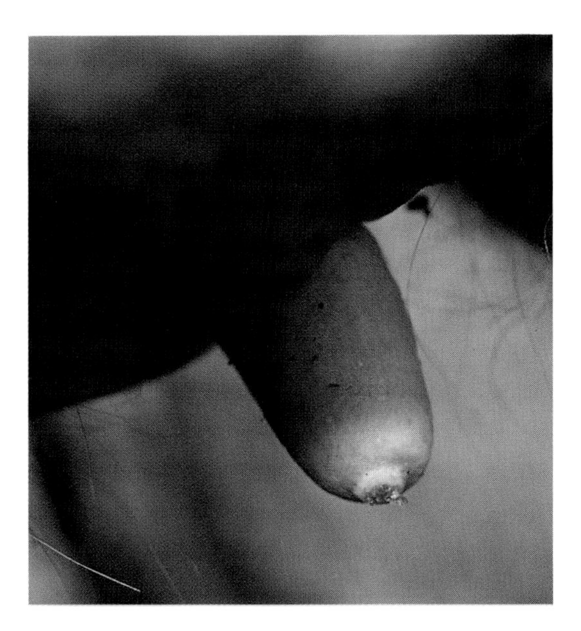

PLATE 14.25 Moderate hyperkeratosis and 'raised' teat orifice: teat score 2.

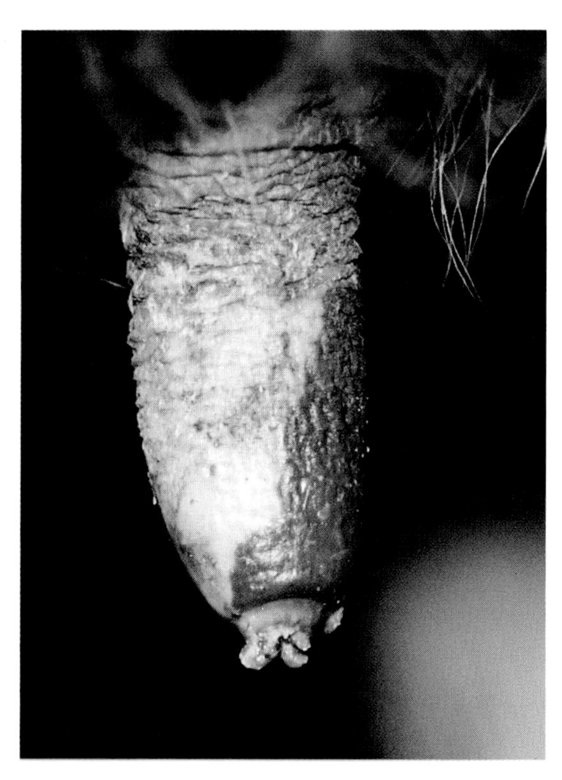

PLATE 14.26 Severe teat-end hyperkeratosis: teat score 4.

Teat-end Haemorrhage

Small haemorrhages, as seen in Plate 14.27, are said to be the result of poor machine function and suboptimal pulsation, leading to inadequate teat massage during milking. This is discussed in more detail on page 65 and demonstrated in Plate 5.27. Plate 14.28 shows a more advanced case. Note the extensive haemorrhage around the teat end, the protrusion of the sphincter and also the haemorrhage at the base of the teat, adjacent to the udder. This is caused by the liner crawling up the teat and pinching it closed at this point. In liners with shields (see page 57) the effective length of the liner may be reduced so much that it fails to compress the teat end fully during the rest phase of the pulsation cycle (see page 53). This leads to poor blood circulation at the end of the teat and can also cause teat-end haemorrhage.

Teat-end Oedema

Swelling and oedema of the teat end, seen especially immediately after the milking unit is removed (Plate 14.29) is particularly common in heifers and freshly calved cows. The swelling and line of flattening of the teat will follow the plane of collapse of the liner, since liners always open and close in the same lateral plane. Hence if the teat in Plate 14.29 was viewed from the side, it would in fact appear thinner, rather than fatter. Compression of the teat may be so severe that a wedge forms across the teat end. In advanced cases this may crack and produce chaps.

Like any other physical damage, teat-end oedema reduces the effectiveness of the various defence mechanisms described on page 17 and consequently increases the risk of mastitis. It is thought that a variety of factors contribute to teat-end oedema.
These include:

- overmilking
- excessively high or fluctuating vacuum

- poor pulsation with inadequate massage phase (see page 65)
- excessively worn liners
- inadequate cluster weights
- poor ACR adjustment, leading to the cluster being pulled off while still under vacuum

In summary it can be seen that there are a variety of teat and udder lesions, some of which are contagious, many of which will lead to an increased risk of mastitis. It is important that the lesion is correctly diagnosed so that remedial action can be taken to ensure that the effect on mastitis can be minimised.

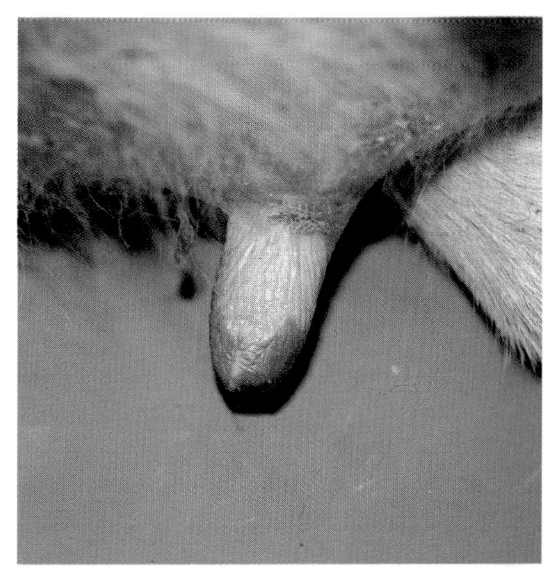

PLATE 14.28 Severe teat-end haemorrhage.

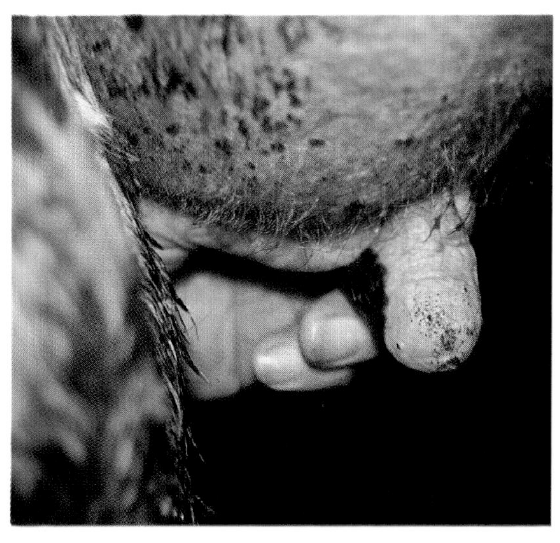

PLATE 14.27 Mild teat-end haemorrhage.

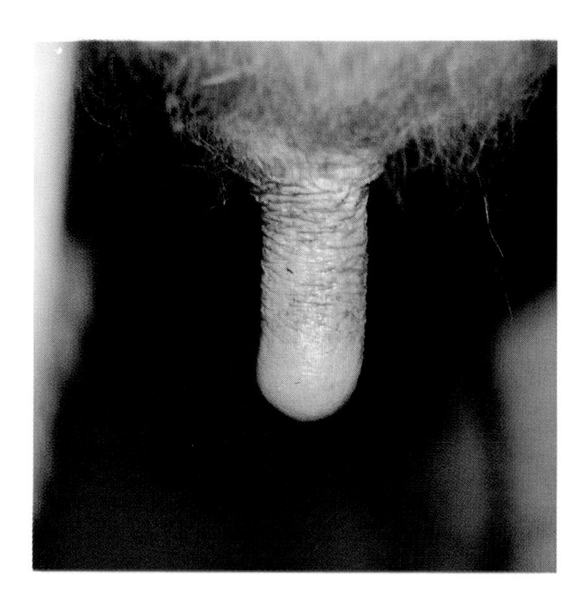

PLATE 14.29 Teat-end oedema: note the swelling at the end of the teat, best seen immediately after removal of the milking machine.

APPENDIX

Liner Life in Days According to Herd and Parlour Size and Assuming a Liner Life of 2,500 Milkings

TWO TIMES A DAY MILKING

Herd size	*No. of milking units*							
	6	*8*	*10*	*12*	*14*	*16*	*18*	*20*
50	150	182	182	182	182	182	182	182
60	125	167	182	182	182	182	182	182
70	107	143	179	182	182	182	182	182
80	94	125	156	182	182	182	182	182
90	83	111	139	167	182	182	182	182
100	75	100	125	150	175	182	182	182
110	68	91	114	136	159	182	182	182
120	62	83	104	125	146	167	182	182
130	58	77	96	115	135	154	173	182
140	54	71	89	107	125	143	161	179
150	50	67	83	100	117	133	150	167
160	47	62	78	94	109	125	141	156
170	44	59	74	88	103	118	132	147
180	42	56	69	83	97	111	125	139
190	39	53	66	79	92	105	118	132
200	38	50	62	75	88	100	112	125
210	36	48	60	71	83	95	107	119
220	34	45	57	68	80	91	102	114
230	33	43	54	65	76	87	98	109
240	31	42	52	62	73	83	94	104
250	30	40	50	60	70	80	90	100
260	29	38	48	58	67	77	87	96
270	28	37	46	56	65	74	83	93
280	27	36	45	54	62	71	80	89
290	26	34	43	52	60	69	78	86
300	25	33	42	50	58	67	75	83

THREE TIMES A DAY MILKING

Herd size	No. of milking units							
	6	*8*	*10*	*12*	*14*	*16*	*18*	*20*
50	100	133	167	–	–	–	–	–
60	83	111	139	167	–	–	–	–
70	71	95	119	143	167	–	–	–
80	62	83	104	125	146	167	182	–
90	56	74	93	111	130	148	167	182
100	50	67	83	100	117	133	150	167
110	45	61	76	91	106	121	136	152
120	42	56	69	83	97	111	125	139
130	38	51	64	77	90	103	115	128
140	36	48	60	71	83	95	107	119
150	33	44	56	67	78	89	100	111
160	31	42	52	62	73	83	94	104
170	29	39	49	59	69	78	88	98
180	28	37	46	56	65	74	83	93
190	26	35	44	53	61	70	79	88
200	25	33	42	50	58	67	75	83
210	24	32	40	48	56	63	71	79
220	23	30	38	45	53	61	68	76
230	22	29	36	43	51	58	65	72
240	21	28	35	42	49	56	62	69
250	20	27	33	40	47	53	60	67
260	19	26	32	38	45	51	58	64
270	19	25	31	37	43	49	56	62
280	18	24	30	36	42	48	54	60
290	17	23	29	34	40	46	52	57
300	17	22	28	33	39	44	50	56

References and Further Reading

BCVA Cattle Practice = Proceedings of the British Cattle Veterinary Association in the journal *Cattle Practice*
J Dairy Res = *Journal of Dairy Research*
J Dairy Sci = *Journal of Dairy Science*
NIRD Tech Bull = *National Institute for Research in Dairying Technical Bulletin.* (NIRD now called Animal Grassland Research Institute)
Proc Brit Mast Conf = Proceedings of the British Mastitis Conference
Proc 3rd Int Mastitis Seminar = Proceedings of the 3rd International Mastitis Seminar
Res Vet Sci = *Research into Veterinary Science*
Vet Rec = *Veterinary Record*

(*1*) *Genus Animal Health* (1994)
(*2*) *International Dairy Federation* (1993)
(*3*) Esslemont (1993), Daisy, University of Reading
(*4*) Philpot, W. N., and Nickerson, S. C. (1991), *Mastitis Counter Attack*, Babeson Bros, USA
(*5*) Harrison, R. D., Reynolds, I. P., and Little, W. (1983), *J Dairy Res*, 50, p. 405
(*6*) Bramley, A. J., Godhino, K. S., and Grindal, R. J. (1981), *J Dairy Res*, 48, p. 379
(*7*) Grindal, R. and Hillerton, J. E. (1991), *J Dairy Res*, 58, p. 263
(*8*) Hill, A. W. (1981), *Res Vet Sci*, 31, p. 107
(*9*) Lee, C. S., Wooding, F. B., and Kemp, P. (1980), *J Dairy Res*, 47, p. 39
(*10*) Hill, A. W. (1981), *Res Vet Sci*, 31, p. 107
(*11*) Adapted from Hill, A. W. (1990), Proc Brit Mast conf, p. 49 and Booth, J. M. (1993), *BCVA Cattle Practice*, 1, p. 125
(*12*) Tyler, J. W. and Baggot, J. D. (1992), *Bovine Medicine*, Blackwell Scientific Publications, Oxford, p. 836
(*13*) Bramley, A. J. (1992), *Bovine Medicine*, Blackwell Scientific Publications, Oxford, p. 289
(*14*) Bramley, A. J. (1980), *Mastitis Control and Herd Management, NIRD Tech Bull*, 4, p. 57
(*15*) Hill A. W. (1992) Proc Mastitis Conf, Glos
(*16*) Hill A. W. (1992) Proc BCVA 1991–92, p. 279
(*17*) Bramley, A. J. (1980), *Mastitis Control and Herd Management, NIRD Tech Bull*, 4, p. 71
(*18*) Hill, A. W. (1992), Proc BCVA 1991–92, p. 281
(*19*) Spencer, S. B. (1990), *The Basics of Vacuum in Milking Systems and Milking Management*, Harrisburg, Pennsylvannia, USA
(*20*) O'Callaghan, E. (1988), Proc *Milking Systems and Milking Management*, Harrisburg, Pennsylvania, USA
(*21*) Neave, F. K., Dodd, F. H., Kingwill, R. G. and Westgaith, D. R. (1969) *J Dairy Sci*, 52, pp. 696–707
(*22*) Neave, F. K. (1971) *The Control of Bovine Mastitis*, ed. Dodd, F. H. and Jackson, E. R., British Cattle Veterinary Association
(*23*) Galton, D. M., Petersson, L. G., and Merrill, W. G. (1988), *J Dairy Sci*, 71, p. 1417
(*24*) Pankey, J. W., Wildman, E. E., Drescher, R. A., Hogan, J. S. (1987) *J Dairy Sci*, 70, p. 862
(*25*) Bramley, A. J. (1980), *Mastitis Control and Herd Management, NIRD Tech Bull*, 4, p. 74

(26) Rendos, Eberhart and Kesler (1975), *J Dairy Sci*, 58, p. 1492
(27) Bramley, A. J. (1992), *Bovine Medicine*, Blackwell Science Publications, Oxford, p. 294
(28) Booth, J. M. (1994), Proc Brit Mastitis Conf
(29) Adapted from Philpot, W. N. (1984), *Veterinary Clinics of North America Food Animal Practice*, 6
(30) Fuhrmann, T., unpublished data
(31) Philpot, W. N., and Nickerson, S. C., *Quality Milk Production and Mastitis Control*, Holstein Association, USA
(32) Adapted from Chommings, K. J. (1990), *Vet Rec*, 115, p. 499
(33) MacKellar, Q. A. (1991), *In Practice*, 13, p. 244
(34) Tyler, J. W., and Baggot, J. D. (1992), *Bovine Medicine*, Blackwell Science Publications, Oxford, p. 836
(35) Owens, W. E., Watts, J. L., Broddie, R. L., and Nickerson, S. C. (1988), *J Dairy Sci*, 71, p. 3143
(36) Sol, J., Sampimon, O. C., Snoep, J. (1995), Proc 3rd Int Mastitis Seminar, Israel, p. 5–68
(37) Booth, J. M. (1980), *In Practice*, 3, p. 102
(38) Meany, W. J. (1992), Proc BCVA 1991–92, p. 21
(39) Meany, W. J. (1992), Proc BCVA 1991–92, p. 211
(40) Schukken, Y. H., Vanvliets, Vandegeer, D. and Grommers, F. J. (1993), *J Dairy Sci*, 76, p. 2925
(41) Meany, W. J. (1994), *Mastitis and Milk Quality*, Moorepark, Ireland
(42) Bouman, M. (1991), Proc BCVA 1990–91, p. 383
(43) Hillerton, J. E. (1988), *In Practice*, 10, p. 131

Further Reading

Andrews, A. H., Blowey, R. W., Boyd, H., and Eddy, R. G. (1992), *Bovine Medicine*, Blackwell Scientific Publications, Oxford
Blowey, R. W. (1988), *A Veterinary Book for Dairy Farmers*, 2nd edn., Farming Press, Ipswich
Bramley, A. J., Dodd, F. H. and Griffin, T. K. (1980), *Mastitis Control and Herd Management*, NIRD Tech Bull, 4
Bramley, A. J., Bramley, J. A., Dodd, F. H., and Mein, G. A. (1992), *Milking Machine and Lactation*, Insight Books, Berkshire

Index

U

FARMING PRESS BOOKS & VIDEOS

Below is a sample of the wide range of agricultural and veterinary books and videos we publish.
For more information or for a free illustrated catalogue of all our publications please contact:

Farming Press Books & Videos
Miller Freeman Professional Ltd
Wharfedale Road, Ipswich IP1 4LG, United Kingdom
Telephone (01473) 241122 Fax (01473) 240501

Cattle Lameness & Hoofcare
ROGER BLOWEY
Common foot diseases and factors responsible for lameness are described in detail and illustrated with specially commissioned drawings. Full details on trimming.

Footcare in Cattle, Hoof Structure and Trimming
(VHS colour video)
Roger Blowey first analyses hoof structure and horn growth using laboratory specimens, then demonstrates trimming.

A Veterinary Book for Dairy Farmers
ROGER BLOWEY
Deals with the full range of cattle and calf ailments, with the emphasis on preventive medicine.

Cattle Ailments EDDIE STRAITON
The recognition and treatment of all the common cattle ailments shown in over 300 photographs.

Calving the Cow and Care of the Calf
EDDIE STRAITON
A highly illustrated manual offering practical, commonsense guidance.

Cattle Behaviour CLIVE PHILLIPS
Describes what cattle do and why – what is normal in young and old, male and female, and what is not.

Cattle Feeding JOHN OWEN
A detailed account of the principles and practices of cattle feeding, including optimal diet formulation.

Farming Press Books & Videos is a division of Miller Freeman Professional Ltd which provides a wide range of media services in agriculture and allied businesses. Among the magazines published by the group are *Arable Farming, Dairy Farmer, Farming News, Pig Farming* and *What's New in Farming*. For a specimen copy of any of these please contact the address above.